Altimeter Rising -
my fifty years in the cockpit

By Alan MacNutt
with Norman Avery

ALTIMETER RISING

Copyright © ALAN MACNUTT, 2000

All rights reserved. No parts of this book may be reproduced, stored in a retrieval system, or transmitted, in any form or by any means, electronic, mechanical, recording, or otherwise without the prior written permission of the author, except for purposes of review, in which passages may be quoted in print, or broadcast on television or radio.

Cataloguing in Publication Information

MacNutt, Alan 1924-

Altimeter Rising —
my fifty years in the cockpit

ISBN 0-9687627-0-0

1. Air pilots — Canada — Biography. I Title.

TL540.M27A3 2000 .. 629.13'092   ..C00-901032-7

Published by:
Alan MacNutt
904 - 3170 Gladwin Road
Abbotsford, B.C. V2T 5T1

Internet: imacnutt@dowco.com
Telephone: (604) 852-3216 Fax: (604) 859-3218

Contents

Lost in a Thunderstorm 1
Learning to Fly in Wartime 4
The Flying School Era 17
The Canadair Days 29
Spartan of Ottawa 37
The Flying Husky 49
Churchill – the Big Smoke 62
One Burning, One Turning 67
The Brain Game 77
Propwash 82
The Kenya Caper 87
The Bradley Experience 95
Survair Ltd. Comes of Age 106
Ferry Flights 114
The Longest Night 124
Incident in Colombia 131
R.O.N. in Santa Maria 135
The Yankee Dollar 141
High Over the Andes 145
The North Sea Find 148
Aerobatics in Morocco 152
The DOT Years 157
Flying with the Tsetse Flies 165
Conair Aviation 169
Bullseye on the Chopper 174
Captain Bob Learns the Trade 179
My Love Affair with the Canso 187
Time to Move on 200

Dedication

For Irene and our three sons, Bob, Jim and Jack – all pilots, all accident free

Special thanks to the following who made this book possible: Norm Avery, author of Whiskey Whiskey Papa, for research, encouragement and reality checks, Jim Turner of PixelGraph Studio in Ottawa for the cover and layout, Dwayne Olson, Pete Mitchell, Norm Spraggs, Doug. McDonnell, and many others over the years.

Manned flight has done more to change the relations of the earth's inhabitants than any other factor of human endeavour. As an awesome power for good and for evil, it has nullified the protection and isolation of traditional frontiers. It has lofted the common man toward hitherto undreamed of destinations, and it has changed forever the very pattern of his daily existence.

Ernest K. Gann

Foreword

Al MacNutt's recollections of life in the cockpit attest to his fascination, bordering on obsession, with aviation. He was ill-equipped when he followed his brothers into the Royal Canadian Air Force during the Second World War. But he persevered among his better-educated colleagues, studied hard and finally conquered the fine art of flying an airplane. You could, perhaps, call it luck that the air force graduated him into a surplus of new pilots, luck that the Royal Navy was just down the street, looking for pilots.

Al's determination to be the best flyer in the business kept him striving to that end. It took him through the rigors of mastering the Navy's Seafire over Britain, then through the rough and tumble of his own flying school. Adversity, tinged with good fortune, landed him in the jet age with Canadair and the RCAF Auxiliary, thence into aerial survey in the Arctic and foreign lands. He went on to test airlines and their pilots, and retired after years as a forest fire bomber pilot in western Canada. It has been a career laced with exciting ventures, adventures and some misadventures over tundra, jungle, desert, mountain and ocean.

Throughout his 50-years in the cockpit, Al had his ups and downs in the business. His memoir, however, will leave you with no doubt that he has landed and taxied to the ramp abundantly rewarded for his stubborn pursuit of excellence in aviation.

Norm Avery

ALAN MACNUTT

The author waits clearance for a mapping survey run in his P38 high over the mountains of Chile.

Introduction

In 1992, my fiftieth year as a pilot, I looked back over my career and considered how fortunate I had been to have had such a long and enjoyable working life. The flyer's life is different. No special background, education, wealth or wisdom distinguishes the pilot Those of us that have been privileged to enjoy the profession have been richly blessed.

Many of the 67 aircraft types I have flown are gone, swept away by the rapid advances of technology even before they could be preserved as museum pieces. Many changes have taken place in aviation in that half century. Speed, efficiency, comfort and safety have improved as we experienced a burgeoning increase in air traffic. Aerial survey showed us, years before satellites, what our world really looks like from above.

Aviation has shrunk the globe.

Alan MacNutt

ALTIMETER RISING

The Canso was a favourite aircraft. It served on survey flights in South America and Africa. This one was used in an ill-fated enterprise transporting fishermen up the Pacific coast. (see "My Love Affair with the Canso"... p.187)

Lost in a Thunderstorm
(or, a rolling snowball gathers a ticket to eternity)

SNOWBALL *n. snow pressed into ball, esp. as missile; anything growing or increasing rapidly, like snowball rolled along ground.*

That's what the dictionary says. But that description fails to include the 'flyer's snowball' which should be defined as follows:

SNOWBALL *(flyer) n. airborne situation in which pilot allows one miscalculation to grow rapidly and compound ensuing problems into an ever-increasing crisis, like a (runaway) snowball rolled along ground.*

----oOo----

The Ottawa Valley and the Gatineau Hills to the north are very beautiful in autumn. The maple trees take on colours that cannot be matched in the forests of any other country. The small towns on both sides of the Ottawa River are quiet, quaint and seemingly oblivious to the political convolutions that take place farther down the river.

One October day in the late 1940s I was flying an Aeronca Champ from North Bay to Ottawa, a distance of about 225 miles.

ALTIMETER RISING

There were storm clouds to the west and north but the weather office assured me that I could outrun them easily. Maybe I didn't take time to explain that the top speed of a Champ is about 80 mph. And I didn't mention that it had no radio that could receive weather updates, or have instruments that would allow me to fly in cloud. Or maybe he miscalculated where the storms were located and where they were going. In any case, I pressed on.

I had a bad case of get-home-itis and felt confident that I could reach Uplands Airport south of Ottawa before dark. It is safe to assume, in retrospect, that I didn't have the cash for a hotel room in North Bay. What I did have was lots of youthful enthusiasm, a highway, a railway, and a big river to follow, and both Arnprior and Carp airports for alternates along the route.

One of the factors in my developing snowball of trouble was the clock's 'fall back' day – the move from Daylight Saving back to Standard time. Even though I had reset my watch in the morning, I had failed to consider the earlier evening into which I was flying. Darkness, to my amazement, came one hour early. At the same time an undercast started to slide in below me, blocking reference to my Ottawa River guide. As the snowball increased in complexity I punched into thick cumulus clouds. Heavy rain pelted the Aeronca. Finally, when I was having great difficulty keeping the aircraft right side up, I became trapped in a wild thunderstorm. Turning back was not an option. I was surrounded. Down below I could only catch brief glimpses of the ground through holes in the low scud. The river, the road and the railway were long gone and my little cotton-covered trainer was getting thrashed around the sky like a rag doll.

Vertical columns inside a thunderstorm rip up and down like high-speed elevators passing each other in a skyscraper. The severe turbulence can toss a 300-ton aircraft around like a toy. My Aeronca weighed about 1,500 pounds fully loaded and I was being jerked and dashed around the sky like a runaway kite. I was afraid to go too low in case I flew into the Gatineau Hills but then I didn't have much to say about my altitude or direction. Flashes of lightning provided some light on my basic instruments and illuminated the occasional patch of earth, but map reading was impossible. The trees all looked the same from the top, and the occasional field flashing by would probably have wrecked the airplane if I attempted a landing.

I was competent to use the radio range system of navigation but, of course, the Champ had no radio, no generator, no battery, and I was now convinced, a pilot with no brains. I clung to the ridiculous hope that I could outfly the storm before I ran out of fuel but I had no idea where I was or where to run. Through sheer good luck I caught a glimpse of the four white and orange towers that marked a radio range station. Beside them was a small grass field. Using the frequent flashes of lightning to mark the flat area, I somehow managed to get the Aeronca onto the ground. I jumped out and grabbed the wing strut and held on for what seemed like hours to keep the aircraft from flying away without me.

Next morning, the storm had passed. I found that I had spent the night at the Radio Range Station at Killaloe, a tiny farm town on the Ontario side of the Ottawa River. The unused field had been an emergency site to accommodate aircraft in distress dating from the earliest airline system. I was about 100 miles short of my destination but surprisingly on track. The airplane seemed none the worse for the night of aerobatics. High-speed hail pellets had riddled the aircraft like a machine gun. It had dented the aluminum cowling but had spared the new cotton fabric on the wings. Older 'tired' fabric, however, would have been shredded – me along with it.

Time by the clock had only changed one hour, but a chastened pilot had aged considerably by the time the trip was continued next day down the scenic waterway to Ottawa. The change of viewpoint from a panicky, out-of-control amateur in a night thunderstorm to a cool dude the next morning, however, showed me the gamut of emotion we experience as pilots.

Flying is indeed a joy, but occasionally we must be prepared for harsh reality. Antoine de Saint Exupery the famed author/pilot says it best: "If a pilot is wrong, then just below the sea of clouds begins eternity."

But I'm getting ahead of myself.

Learning to Fly in Wartime

From the humble Tiger Moth to the elegant Seafire

From Tiger Moth to Boeing 747 – from Link trainer to flight simulator. To have been there when all the skyrocketing progress happened has been a privilege pilots of the future will never experience from the same ringside seat. Technology has become too complex, imposing progress that will outrun most fun-seeking pilots. How will the pilot of the future be trained – by man or machine? I admired my teachers in school and revered my wartime flying instructors. As a consequence, I tend to lean towards the human element, a result of my enduring desire to instruct. Some of my most cherished moments were those in which I sent students on their first solo or checked junior pilots out on sophisticated airplanes that would challenge their budding talents. I could watch them grow in responsibility, experience, confidence and character.

The motto of our flying school in the late 1940s was *'We promote **SAFE** flying'*. The school went broke, but many of our students went on to major achievements in the aviation world. At one time, five of my students became chief pilots of various flying companies. Best of all, my wife and three sons all became excellent pilots. This must be success. In retrospect, it might have been advisable for our school to have adopted the motto – '*We promote **PROFITABLE** flying.*'

Our son operates his airline jet at a standard of safety and passenger comfort that was inconceivable when I learned to fly. It is with some pride that I watch 'The Captain' mount his Boeing 737 for his routine flights. Relaxed and comfortable in the main cabins, more than 100 passengers sit confident in his estimate that they will arrive in Halifax, Toronto, Prince George or Dallas on schedule. I am proud to have helped launch him on a career he enjoys so much. Our other two sons are both pilots and air traffic controllers. They are much more knowledgeable than I am about the new procedures, techniques and technology that have advanced aviation into the 21st century.

I finished my commercial aviation career in an aircraft designed in 1942, the year I started flying. The beautiful old Douglas A26 had bullet holes in the perspex and on the armour plating, souvenirs of a colourful past, notably in Korea. During my last several years of flying, it provided a fast, stable, and dependable platform from which to bomb forest fires in Western Canada. The airplane and me, in our advanced years, were both content to accept a modest role in aerial fire protection and reflect on our memories of more exciting days in exotic sounding places all over the globe. I made a point of being very kind to the old airplane and it returned the favour. The A26 was developed from an earlier model called the Marauder, a bomber that had some nasty habits. Douglas first flew the A26 in 1944, long ago in the life of an aircraft but with good maintenance it provided decades of service. Like every aircraft type, the A26 has a personality. Like automobiles or horses, none is perfect. But we can enjoy many characteristics and use the bad habits for something to complain about.

My introduction to flying was not as well planned as that of most pilots whose biographies I have read. I did not build model airplanes nor did I watch barnstormers from the safe side of the fence. My two brothers had joined the RCAF and gone overseas early in the war. It seemed the thing for me to do, too, when I was old enough. It appeared much more exciting than raising potatoes, silver foxes, and purebred Holsteins, my ancestors' occupation for the previous six generations on Prince Edward Island. Fortunately for me there was a shortage of pilots when I enlisted and the recruiting officer had a quota to fill.

ALTIMETER RISING

My first flying adventures were on the Tiger Moth, the elementary trainer of the RCAF. I do not intend to bore you with the details of all the 65 aircraft types I have flown over the years, but the Tiger Moth was different. Designed in Britain by de Havilland, the biplane was mass produced in Canada during the first years of World War II to serve as an elementary single engine trainer. The Fleet Finch was also used and joined later by the Fairchild Cornell which looked and flew more like a military aircraft should.

At one point it was decided to send prospective pilot trainees to a preliminary flight training unit before ground school, to see if they could even qualify as pilots. It seems many had been recruited, sent to long, expensive courses on engines, theory of flight, and navigation. Then, when it came time for actual flying, the student was found incapable, unwilling to try or so scared that he was always airsick. This was a waste of nearly a year ground training. Trainees when recruited would now be sent to an aptitude (or attitude?) course to see if they were indeed potential pilot material. In my case this involved about eight hours dual in a Tiger Moth. We were not expected to solo, just to show that we could manipulate the controls, were not afraid to be airborne and show some ability to apply what we had been taught yesterday.

Learning to fly the mighty Tiger gave me some problems. Turns, aerobatics, take-off and formation flying presented no insurmountable challenge. However, when it came time to land I had a habit of looking straight down instead of ahead. All I'd see was a blur of grass rushing past. I would have no idea how high I was and would either whump into the ground or level off too high and drop like a rock when the wings stalled. Fortunately, my instructors were patient. Flying Officer Redpath found the problem. When he told me to look ahead of the airplane on the left side, it all came into focus and my career in aviation passed another important hurdle.

To me this training was enjoyable and my first escape from CT, the dreaded fear of every student pilot — an order to cease training. Then came months of ground school with CT opportunities at every critical exam. Finally, it was off to

TWO VIEWS OF THE WARRIOR – as an 18-year-old RCAF trainee in 1942 to a seasoned Seafire pilot in the Royal Navy's Fleet Air Arm in England near the end of the war.

One way to survive disaster in a Tiger was to flat spin it down through cloud. The aircraft got well bent, but the pilot usually walked away.

The elegant Seafire did not offer a similar escape but the well-trained pilot had a better chance by using his extensive skills.

ALTIMETER RISING

Elementary Flying School in Oshawa, Ontario, to master the Moth. My RCAF log book shows another 10 hours before that joyful date.

When I finally got the slap on the back that cleared me to fly on my own, my pulse quickened. A few butterflies did a preliminary circuit in my stomach. But I wasn't going to hesitate for a second in case my instructor changed his mind. I had no doubts about my first solo. I taxied out, glancing back to see if anybody was paying any special attention. I checked the wind sock and pointed the aircraft directly into the wind. Little did I appreciate at the time that the green light from the tower was to be such a pivotal point in my life. I opened the throttle and began my run down the grass infield, concentrating as I never had before. The lighter load in the Tiger got me airborne more quickly than during training, but not really quickly enough to suit me. I was in a hurry to get the job done. My task was to fly a single circuit around the airport and land in one piece. And I undertook the assignment as if it were an order to attack the enemy. I resented the area restriction but settled for an uneventful circuit and a creditable landing. Ten minutes of my life had passed but my destiny was changed; farming was history. I could do it. Lindberg watch out!

I taxied back into line and jumped from the aircraft with weakened knees. Handshakes and more back slaps confirmed my right of passage. I was a flyer. But I had no idea there was so much more to learn about flying. Fifty years later I was still learning, and technology was advancing at such a pace that what I learned a decade before was obsolete, often incorrect and dangerous. But of course I didn't know that then, I only knew I wanted to fly.

The early theory was to take off normally then fly down the field a few feet above the ground to gain as much speed as possible then pull up at the fence and fly away. Speed was deemed to be our friend; height was less important. You need only stand at the airport fence nowadays and watch the jets rotate and go immediately into a steep climb, to know the speed theory was incorrect.

To get back to the Tiger Moth at Oshawa, I thought I knew all about aviation after that first solo trip. I had difficulty understanding why my instructor gave me two more hours dual just to make me realize that there was more to learn. I had no idea how much.

The Tiger Moth was powered by a four-cylinder Gypsy Moth engine turning a two-bladed wooden propeller. To give reasonable visibility over the front for the trainees, the engine was installed upside down with the crankcase on the top, the cylinders and heads on the bottom, and being British, the propeller turned the wrong way — counter-clockwise. Due to our Canadian climate, a sliding coupe top was added to keep out the snow and ice. Another modification, made necessary because many of the hastily built training airports were rough grass fields, was a tail wheel. We threw away the tailskid used on the British models. There were no heaters and the instructor, normally flying from the front seat, had to communicate with his student through a hollow 'Gosport Tube' with a mouthpiece on each end. It fed into the other pilot's leather helmet. Compared to the sleek Cessna and Piper trainers of today, it was a beast to fly, but had the advantage that when mastered, the step up to a bigger aircraft seemed relatively easy.

One more comment on the 'Yellow Perils' (all trainers were painted yellow); they loved to spin. We were taught that if we ever were so foolish as to get caught out flying solo and end up in cloud, lost and running out of fuel, we were to put the airplane into a spin. The Tiger Moth's spin was flat. The wide circle it would perform down through the cloud could be corrected once we could see the ground. We were to stop the spin immediately and land somewhere. Many lives were saved this way.

In the event there was not enough distance between the cloud base where we became 'visual' and where we could stop the spinning, the aircraft would always hit the ground nearly flat. This would tear off the undercarriage, demolish the fuselage and collapse the top wing. But the pilot could usually walk away. To descend through the clouds with few instruments and little knowledge of how to use them, we were almost certain to wind up in a high-speed spiral dive which usually proved fatal.

ALTIMETER RISING

Those of us who completed our elementary training went on to the next stage on faster, more sophisticated, airplanes. My luck was to go to Service Flying Training School at Uplands airport in Ottawa to train on Harvards. I then went 15 miles west to Gunnery Training School at Carp. After passing my Wings test, I was hooked. I never wanted to do anything else but fly. A total of 133,000 pilots were trained all across Canada along with navigators, observers, air gunners, and mechanics for the Commonwealth countries.

Evidently, with my new status as sergeant-pilot, I was getting soft. I was promptly sent off to a physical fitness commando course at Trois Rivières. Despite the aches and pains of my commando transformation, I managed to meet a striking blonde who became, along with flying, the shared love of my life. Finally, battle ready, I arrived at Halifax, where it was planned to ship the new pilots overseas to replenish squadrons whose pilots had been lost or tour-expired. However, the program had changed. The air war had been won and the British Commonwealth Air Training Plan had been so successful that a surplus of pilots had been graduated. Instead of being relieved that we were not going to be fodder for the Luftwaffe, we felt cheated, deprived of what we felt was our right to become heroes in the air war. Worst of all, there was the very real possibility that our flying careers were over when we had just tasted the joys of being pilots.

At this point, with the war apparently favouring our side, I had the option of taking a discharge as a redundant pilot. The other option was to join the British Navy Fleet Air Arm and try to continue my new love of flying. This career decision took about 30 seconds — my first point of no return — and a decision I have never regretted.

Casualties had been high and replacements were badly needed. Of our new group, 10 Canadians volunteered. We went a few blocks down Barrington Street, joined the Royal Navy (VR – for Volunteer Reserve) Fleet Air Arm. In one week, were on our way to fame but not much fortune in England in what smelled like a cattle ship. It was the Queen Mary, but as a troopship the traditional luxuries were hard to find.

Canadians, during the war, were much more highly paid than the British, but paid much less than the Americans of comparable rank. On arrival in the UK our lofty Canadian pay of two dollars a day was reduced to three shillings (approximately 66 cents Canadian). Our social life, needless to say, was restricted. Food was in short supply. Our fare was mostly rabbit and brussels sprouts, the latter only one of the many by-products of war. As servicemen we were pampered; most civilians, including children, had far less. Some high-spirited young Canadians would occasionally raid the wardroom galley and liberate a can or two of Spam after a mess party, but I don't remember ever being caught, so I could not have been among them!

When our duplicate training records finally arrived and our credentials were established, our group was commissioned as Sub-Lieutenant (A)s. We started training on operational aircraft and after two short refresher courses, we were assigned to the most beautiful aircraft of them all - the Seafire. The Air Force version was the Spitfire. The Royal Navy added a few modifications such as a flap selector and a tail hook to pick up the restraining cable on carrier landings.

To get us ready for Seafires, our conversion courses were in dual control Miles Master aircraft. The Seafire had only one seat and so no dual training on the type was possible. The Miles Master was a tandem-seated aircraft with a big piston engine. The purpose was to teach us to land when we couldn't see out the front. The Seafire had a long nose and you must land looking out the side. When the tail came down all you could see ahead was cowling.

On my first trip out, trying to taxi the Master from the back seat, I was told to raise the seat as high as possible and leave the coupe top up so that I could see where I was going on the ground. The Chief Petty Officer instructor in the front seat, neglected to tell me — a brash colonial — to be sure and lower the seat before I closed the coupe top for take-off. When I was ready to go I lowered the spring-loaded brute of a canopy. It crashed down on my head and knocked me out on the runway before I got to do my first take-off. The instructor, who no doubt kept a score card on his 'kills', appeared unimpressed.

ALTIMETER RISING

Phase 2 of our training was a refresher course on Harvards for gunnery training before the big step to Seafires. I made one solo trip in the early morning and when I came back, exercises completed, I saw the rest of my peer group sitting out on chairs on the grass in front of the flight shack. It seemed a good idea for a hot shot pilot like myself to land really close to my pals on the grass infield, where they were enjoying tea. Most wartime strips, except for the larger bomber bases, were grass. The grass was wet with dew and braking action was zero and Our Hero had some pressure on the toe brakes. As I watched for the favourable reaction of my buddies to my beautiful landing, I missed noting the narrow paved footpath that led up to the hut. In plain view of the spectators, the three wheels of the Harvard hit the grass for a lovely touchdown. When I crossed the path, the brakes locked and the Harvard went from a three-pointer tail down, to a wheelie, to a three-pointer nose down, with the propeller chewing into the ground to serve as the forward point. The humiliation of sitting there on my nose in front of the whole station, all my 20 years and 250 hours flying time hanging out, was something I won't forget.

The spectacular landing did not interfere with my progress to the Seafire. In peacetime pilots move up by degrees in the sophistication of the airplanes to which they are entrusted. In wartime we were taken from a basic trainer to a sleek powerhouse that flew four times as fast, four times as high. The Seafires had complex radios with four channels (wow!), oxygen masks, hydraulics to raise and lower gear, and complicated instruments. There were no flight simulators for various types, no comprehensive manuals or briefings, and, in the Seafires, no dual instruction.

My flight leader crouched on the wing on my first Seafire solo, showed me how to start it while two 'erks' sat on the tail to ensure that I didn't apply too much power and put it up on its nose. Then all three disappeared and I was 'Captain-Qualified'.

As we had been watching the course ahead of us 'check out' we were well aware that the greatest problem for a beginner was handling the throttle to the huge Rolls Royce Merlin. Abrupt throttle usage on the ground would tip the tail up and the prop would hit the ground. Abrupt use of the throttle on take-off would cause a big yaw as the vertical tail could not control direction at low speeds. Most dangerous of all was a burst of throttle to correct a poor landing and avoid the next bounce. The sudden surge of torque at low speed would cause the little airplane to try and turn the opposite direction as the prop. We had watched several of our friends add a lot of power after a minor bounce (as we were taught to do in training) then roll over on their back and burn.

Instructors learned to tell their students to treat airplanes gently like you would a lady, no abrupt maneuvers near the ground and great things were in store. Sound advice!

After a few hours solo to become familiar with the hot fighter, we were taken out daily by an instructor to practice aerobatics, low flying, and dogfighting (using a camera instead of bullets to record our strikes). Our instructor would take us on cross-country sweeps, usually in poor visibility. Cruising about 300 miles per hour, he would make us fly at treetop level. When we came to a power line we had to go below it and dodge industrial smokestacks by rolling up on a wingtip as we passed. This was all to prepare us for staying under the German radar. It was very exciting for this farm boy.

Looking back at this phase of my flying, I do not think I fully appreciated the fact that as a very young inexperienced pilot, I had had the opportunity to fly one of the most famous aircraft in the world. I could feel its power and performance. Its eye appeal was obvious, but I was unaware that thousands of pilots from that era on would have given their eye teeth to fly that beautiful aircraft, which with the Hurricane, was such a factor in British history. I guess I was too young to appreciate the magnificent lines, the acceleration and climb when you pushed the throttle slowly to take-off power. The manoeuvrability and speed range from landings at under one hundred miles per hour to cruise at about 400 mph were awesome. By today's standards, the airplane was tiny. One writer credits its pure lines to the fact that it was designed by one man, Reginald Mitchell, and not, like most modern aircraft, by a committee.

ALTIMETER RISING

The RNVR Fleet Air Arm was fun. Morale was great, even with V2s falling on London. Unlike the common perception of the stuffy Navy, the only stuffy naval people I met in two years in Britain were the Canadian Navy Officers who seemed to think they had to adopt a British accent and emulate Nelson. Of course, a few of the permanent Royal Naval Officers treated us with some contempt. I did not make much of a contribution to winning the war, but those two years in Europe were probably the best two years of my life. 'War is hell' is a favoured quote, but in my case at least it took a green, poorly-educated farm boy, and started him on a career flying high performance state-of-the-art airplanes. From that time on, I never lost the desire to fly and to steer others towards safe flying habits. Unfortunately, later in my civilian career, several of my ex-bosses did not share this vision, which resulted in some very abrupt career changes. In civil aviation, profit is the great motivator - it means survival of the company, an objective often attained through payroll reduction.

One weekend, between VE Day and VJ Day, I wangled a short leave on the Continent and, having no money, I took my Sally Ann (Salvation Army) one-carton quota of cigarettes and enjoyed top notch food and accommodation in Brussels, using cigarettes for trade negotiations. I did not smoke and only claimed my quota occasionally when needed for barter. I recall writing home extolling the pleasures of good food, music, dancing again with girls wearing silk stockings and pretty dresses. I'm sure my mother was shocked that her baby was tasting such worldly pleasures.

Obviously there were other joys than flying Seafires, but there were not many happy faces back in Britain. Outside the periphery of the American Bases, silk stockings and chocolate bars were unheard of. As a footnote to my two years in Britain on wartime rations, I will never, but **never**, eat brussels sprouts or rabbit again!

After VE Day, the war wound down quickly. The Navy, realizing it would have much work in the Pacific adapted our operational training to what they expected against the Japanese. Only months later came VJ Day and our group still had not had an opportunity to be heroes. We had not fired or been fired at in anger. Patrols over the English Channel were halted and, apart from ferrying some Spitfires and Seafires back to their various storage bases around Britain, our services were no longer required.

Britain was not a pretty place in 1945. Finally, in 1946, we realized it was time to grow up and face the real world, and so we requested discharge and were all shipped home via New York in some style - officers, if not yet gentlemen or heroes.

Our only option left in order to continue flying would have been to transfer to the 'regular' Navy. That prospect was daunting as the Royal Navy officers we had met did not seem like happy people and regarded us 'wavy' Navy types as pilot upstarts, helpful in wartime, but a bother you know, and all Colonials at that! My career as a pilot was in jeopardy, and as I had no knowledge of any opportunities to fly in Canada, I could only wait until I got home and hope the flying jobs, if any, had not all been taken.

It was a big transition for me to fly high-performance aircraft in a foreign country before I was twenty. At each selection board I was afraid I would be washed out. I had no knowledge of science and most of my classmates were from high school or college and had big town sophistication and street savvy. It took many nights of studying in the washrooms - lights were out everywhere else - but I squeaked through the ground school exams and was on a par with the rest for the flying portion. In any case I had learned to fly, and as it seemed much better than working for a living I've been at it ever since, in various permutations. After the first few years, young civilian pilots have less fear of starvation and, if things go really bad, some can escape to the RCAF or join Transport Canada as inspectors. But getting that first paying job can be difficult for new pilots, and I can only advise young pilots that if you have any friends in the business, use them. Fancy resumés and logbooks do little to impress a veteran chief pilot in general aviation. He wants to know where you came from and who will recommend you. As you go father north, the chief pilots ask fewer questions.

At war's end there were thousands of pilots all with the same idea. Most of them had many more hours than me, and much of it on two and four-engine bombers. Their experience gave them a much better chance of being picked up by an airline. Some went to South America to help start small airlines and train local pilots. Some joined other overseas flying companies, such as KLM, but most who had been in serious combat wanted to forget the war and airplanes. I desperately wanted to continue to fly. I got a small loan from my dad, bought shares in a small local flying school in P.E.I., called Paul's Flying Service, which any businessman could have predicted was doomed to fail. A much wiser decision was to pursue the lovely blonde lady I had met in Trois Rivières before going overseas and persuade her to become my wife, in which I was successful. My love for flying and for my wife, continued throughout the years, but often she has questioned which was predominant. As a pilot must spend much time away from home, this puts strain on a marriage. When a young woman has three sons to raise, and her husband is flying somewhere in Africa or South America or the Arctic, she needs great intestinal fortitude to hang in there. The upside, however, is we have enjoyed a hundred or more happy reunions, which in some way helps to compensate for the many lonely nights apart.

The share in the flying school kept my dream of flying open. I promptly obtained my Commercial Licence and Instructor's Rating, and started teaching people to fly. We were impoverished, but I was still flying, gaining experience, licences and endorsements, and setting the ground work for what has been a very happy and rewarding career.

ALAN MACNUTT

The Flying School Era

We flew for food

It did not take long to realize that our share in Paul's Flying Service in Charlottetown, P.E.I., was not going to produce a living. And so, while our equity still included a nearly-new two-seat Aeronca Champion, my wife and I put our meagre belongings into the tiny baggage compartment and started the long trek westward. It was a migration that began in 1947, with stops of several years in many cities searching for the ideal place to call home. Finally it ended in the Vancouver area, which is by far the nicest place to live in Canada, and, based on our years overseas, the nicest place in the world.

Our first stop was to visit Irene's parents in Trois Rivières, Quebec. To our delight, there were opportunities in aviation there at the time. We stopped long enough to start a family, Irene to learn to fly, and for us to establish the pattern for our life together.

My mentor in those days was Jack Parlee, a fellow Maritimer who ran Cap Airways in Cap de la Madeleine, Quebec. Cap Airways started as a civil continuation of the elementary flying school that died with the British Commonwealth Air Training Plan at the war's end. In the neighbourhood there were lots of people with lots of money wanting to learn to fly. Cap Airways, however, had a lucrative charter business flying wealthy sportsmen, playboys, hunters and fishermen into the clubs and bush camps. Jack was glad to let me look after the flying school. He helped me get a license from the Air Transport Board. It was registered as Champlain Air Services, for operations apart from Cap Airways.

ALTIMETER RISING

Cap Airways had a Norseman (CF-AYO) the Noorduyn prototype No. 1 with a Wright engine. They also operated four-place Stinson Voyagers and a Stinson Station Wagon on floats from the St. Lawrence River in summer and on skis in winter from Cap de la Madeleine. Skip Lemarier, who had a colourful career before 'retiring' to DOT, was on the pilot staff. Jack Parlee had worked on Stranraer flying boats during the war and instructed at the wartime elementary flying school. Being a Maritimer, he had always wanted a boat and, after retirement, found one he could afford in Newfoundland. He planned to convert it into a charter boat in the Bahamas. While sailing his new purchase from Newfoundland to Maine for a refit, he and his inexperienced crew hit a bad Atlantic storm and were lost. He and his wife Jean were our good friends and for years we four were all poor and happy together. Pilots and mariners have much in common and have a professional respect for each other's avocation.

Jim Williams was another pilot for Cap Airways at this time, and we became good friends. The son of a paper mill executive, Jim later joined Air Canada, but left for medical reasons. With the typical resilience of a bush pilot, he bought a good camera, two motor scooters, and circled the globe with his wife, taking pictures. On his return, he was welcomed into the expanding world of television and had an interesting life as a movie producer for TV in Ottawa. He wrote an excellent book entitled 'The Plan,' about the British Commonwealth Air Training Plan that trained aircrews and ground support personnel for World War II. It is highly recommended reading.

We worked very hard at our flying school - 364 days a year. Starting a fleet of Tiger Moths' engines on a cold winter morning in a snow bank was no easy chore. Teaching the science of aviation on the ground or in the air, in a noisy airplane, in a different language, is difficult. We took over the flying school in Cap de la Madeleine, across the St. Maurice River from Trois Rivières, and later opened a satellite base in Sherbrooke, in the Eastern Townships of Quebec. Irene was from the area and spoke French as easily as English and when she learned to fly it was obvious she would be an excellent bilingual instructor. One obnoxious government inspector, however, changed her mind about this and when she got the nesting instinct and bore three handsome sons in a row, it was obvious her flying career was, at best, delayed.

The aircraft we used for training and charter included the Aeronca Champ 7AC - an excellent, sturdy, economical trainer; a rented Aeronca 11AC Chief; two Stinson 105's - an economic three-seater (one rented); several Tiger Moths; Taylorcraft BC12D tandem seating (looked like a poor man's Fleet Canuck); a rented 4-place Stinson Voyager, state-of-the-art in those days; some Fleet Finches with radial Kinner 5B engines, and later two PA12 three-place Super Cruisers, ideal aircraft for charter or barnstorming in winter. We were always buying and selling old airplanes and trying to build up a fleet suitable for our business. Parts for the war surplus aircraft were in good supply.

Often a student learned to fly on a certain aircraft and would then wish to buy it when he had his licence. As the Americans had a plan to teach their war veterans to fly, there was a big supply of cheap trainers available just across the border in New York State and New Hampshire and Vermont. Often on a weekend I would hitchhike down, buy an old trainer and fly it home. The Customs, relicensing and transfer fees were small, and before our government negotiated free trade such purchases were relatively simple.

Irene had learned to fly in Charlottetown, in 1946, shortly after we were married. She felt if we were going to live together, it would have to be at the airport, so instead of being an airport widow, she became a pilot. One winter day, operating on skis, she was taking her mother to visit her sisters in Sherbrooke. Her mother enjoyed flying and was proud to have her only daughter as her personal executive pilot. For some reason, the Aeronca got off course after crossing the St. Lawrence River and drifted off to the west of track. Let's say the visibility was bad. Irene saw some big hills ahead, but they were not the right shape and she couldn't relate them to her map. Wisely, while she still had lots of fuel left, she found a farmer's long level field and landed safely. They walked to a large industrial establishment nearby. They did not expect the reception that greeted their arrival. The security force was somewhat upset that two strange women had flown into their property undetected. She had landed at Beloeil, south of Montreal, an area noted for apples, wines, and scenery. The well-guarded plant whose field she had chosen produced explosives.

Nearly all of my students were French-Canadian, and I was working in a French speaking environment all day. With the help of Irene and her brother, Lloyd Solly, I gained some facility in the language as it applied to aviation. It surprised me to learn that many of the English words that applied to airplanes, like empennage, aileron and fuselage were actually French words. My students were very tolerant of my brutal use of French verbs, but were helpful. For this I have always felt grateful to the Quebecois. I understand their worry about losing their language and their culture. However, such sympathies are best not expressed West of Winnipeg. French is now used along with English for radio communication air to ground and this upsets some people and can lead to misunderstandings, affecting both politics and air safety.

Maurice Baribeau, a senior officer in the Department of Transport in Montreal District, once gave me some good advice. He was perfectly bilingual and a real gentleman. When I was telling him how hard it was for me to think in French and explain how to fly in another language, he put my mind at ease. He told me that students don't really learn from what you *tell* them but from what you *show* them. He explained that if I spoke beautiful French to my

IRENE'S BIG SMILE celebrates her success in passing her float endorsement. Her bilingualism helped keep the flying school afloat.

students and flew poorly myself, I'd turn out poor pilots. But if I could explain the basics, or even keep quiet, and *do* everything properly, they would learn quickly. The talking could be done on the ground.

    Benoit Rivard, a carpenter, could not speak one word of *anglais* when he came to us; yet he graduated with honours and went on to become Chief Pilot of Eastern Provincial Airlines, a major airline. Jean-Marie Pitre, a papermaker, had no knowledge of English but soon found he could understand my English better than my attempts at French. He joined Trans-Canada Airlines and had a brilliant career in aviation. Leo Lejeune, a handsome lad from La Tuque, was the exception. He spoke beautiful English and after finishing his commercial pilot's license with us, has flown ever since. He ultimately took charge of all aircraft operations involving aerial fire control for the Province of Quebec. Many other young men who learned to fly on Tiger Moths soon found after graduation that almost anything else was easier to fly. At one point I took great pride in being able to say truthfully that five of my graduates became chief pilots for some aviation concern.

    To supplement our meagre income we bought a war surplus aerial camera and took birdseye shots of all the local hotels, paper mills, estates and businesses. We would flog the results on bad weather days and in winter when business was quiet. Sundays we reserved for barnstorming. Each Sunday, especially in winter, when all the aircraft were ski-equipped, we would pick a rural town and arrive overhead with two aircraft, extra fuel, and a bilingual ticket seller. We would circle the parish church until the congregation came from Mass, usually about 11 a.m. By then we would have selected a field near the church and we'd land and approach the village priest directly. He would be taken for a free ride around his parish. By the time the flight landed, we would have a line-up of passengers for *"un tour de ville, trois piastres chaque."* (That's three bucks a flight) These farm boys and girls were a delight to fly. They were all dressed in Sunday finery and, as there was really nowhere to spend it in rural Quebec in winter, had lots of spare cash. Each passenger would come back and regale his timid friends with stories of how we 'looped the loop' and other exciting tales of derring-do. They always seemed to feel they got their three dollars worth.

The Stinson 105 was a great aircraft for these flights. It was super gentle and slow. Its 75 h.p. engine was not too noisy. It did not take off quickly from sticky snow, especially with three sizeable passengers aboard. Until the fuel burned down and reduced the take off weight, the first few take-offs were judged very carefully to get the maximum run. If you didn't have enough airspeed to fly by the time you reached the first fence, you lowered half flap with the right hand, pulled back with the left hand and eased it over the fence, then, flap up, you let it pick up more speed in the next field. If this created anxiety, the pilot was close enough to the customers to provide reassurance. The 105 was one of the most forgiving aircraft I have ever flown. Tiger Moths, on the other hand, were very poor for barnstorming. It was hell to get them started in winter. You had difficulty seeing posts and people on the ground when taxiing, especially on skis. It was hard to get passengers in and out and it burned seven gallons an hour as opposed to 3.5 for the Stinson 105. At 35 cents a gallon that was big money.

The Tiger Moth was powered by a Gypsy Major engine mounted in an inverted 4-cylinder, in-line configuration. Only one of the two magnetos had an impulse for starting. There were no starters or electrical equipment and the impulse often stuck on the good magneto. The remedy: get a metal rod - the spare control column normally - and poke it up into the engine nacelle to bang the magneto so the impulse would release. Even then, on skis it was a tricky business swinging the propeller by hand. When the engine started, the person in the cockpit had no way of keeping the aircraft from sliding ahead and running you down. Swinging the prop from the rear was an option, but it was not easy.

At that time dual instruction cost $8 to $10 per hour and solo $6 to $8 (fuel included) depending on the aircraft. The instructor got $2 per flying hour, but no basic salary. To assist general aviation and provide a backlog of pilots, the government paid each graduate pilot $100 cash. The school also was paid $100 when training to private pilot standard. As the total cost of a private license was only about $400, the cost to the student was often less than $300. Sadly, many took the course, flew a couple of times solo or with a friend, and never flew again. They seemed to want to prove they could fly, but obviously did not get the enjoyment out of flying that has captured so many people. For this reason, many small schools like ours prospered as long as there were new people wanting to learn to fly. However, when we ran out of new students there was not the demand to rent our aircraft or buy their own, that would have kept the schools solvent.

The flying clubs fared better. Because of their close relationship with the government through the British Commonwealth Air Training Plan, they were favoured with free use of hangars and other treatment that the commercial schools could not share. In later years, the difference between a flying school and a flying club became less and less. Some cynics say that the clubs survived because they had a bar to serve drinks (not always after flying finished for the day), and the bar profits helped their bottom line.

The Department of Transport became concerned that I could not afford a full time maintenance engineer to service our fleet. I would do all the necessary work on the aircraft, then take the logbooks (two per aircraft - one for the engines, one for the airframe) downtown to a retired engineer who would sign them out for the week at a cost of $5.00 cash per book. As a result, when an aircraft went to Art Jervis in Lac la Tortue for annual renewal of its Certificate of Airworthiness, there was a lot of major work to do and Art usually had to wait for his money. Art was a Britisher, a man with a dry sense of humour and very helpful. I helped teach his daughter, Peggy, to fly a Tiger Moth.

DOT decided to let me write my Aircraft Engineer's License, called and A and C, and be responsible for my own daily maintenance. We had four 'practical' tests in those days in addition to the written theory exam, which was elementary. The practical tests, as I recall, were (a) splice a piece of steel control cable, (b) rivet two pieces of aluminum alloy, (c) do a dovetail splice on two pieces of wood, and (d) I can't recall. When my wood splice, full of casein glue, was tested in regional inspector Bob Noury's office in Dorval the splice held and his chair leg brace broke. I had my licence! I have never pretended to be a good engineer but my licence is still valid and it has helped me get jobs a few times when jobs were scarce in general aviation. Like automobiles, light airplanes have become increasingly technical and require on-going specialized training for engineers to remain current in the maintenance field today.

After a long day instructing, hopefully with an occasional break for a short charter trip when I would not have to talk all the time, there would be ground school at night. A training manual, 'From The Ground Up' by Sandy MacDonald, came out in those days and was a great help (it cost a dollar then). It still had to be translated into French for local use, and I'm afraid I did not do the job very well. There are probably thousands of books available on how to fly today - then we had only the one, apart from the Air Force manuals, which were of little value to a civilian.

Highlights that come to mind would be the annual arrival of Clare Leavens in his autogyro on the way east to Quebec City. Carl Millard would occasionally drop in to try and sell us a new Beechcraft Bonanza. What a hope! Ever since then I have wanted to be able to afford a Bonanza. Carl would always bring along one or two of his beautiful stewardesses from Trans-Canada Airlines where he worked in those days. When we saw Carl with his new Beech and beautiful girls, moonlighting in aircraft sales from a captain's job, it was obvious all pilots were not created equal.

Jack, Bob and Don Scholfield, were starting Laurentide Aviation out of Cartierville Airport in Montreal at that time, with an Aeronca, a tired Cessna 120 and a used Ford car. They parlayed this into the top school in Canada and started thousands of Canada's commercial pilots on their way to successful careers. The school still operates today at Cedars, just south of Montreal. What a magnificent contribution they made to aviation in Canada. They had the Cessna dealership and sold many aircraft to their former students and to commercial air carriers. Their brother-in-law, Bill Murray, a neighbour and friend, continued to fly after 'retiring' at age 60 many years ago.

Instructing gets boring after the first year and, like a cab driver, you always have to be available when the student wants to fly - not when you feel like working. As well, you have to be up and alert, mentally and physically, or the student gets cheated. Instructing is so very important in the flying business, but unfortunately it is often left to the youngest and most inexperienced. Only a few people have stayed with instructing as a career and a few of them should have retired sooner. It must be a little like motherhood - after you have raised some, you say 'enough is enough'. It is hard not to become jaded and difficult to remain fresh and enthusiastic after you have told a dull student the same thing at least a hundred times and he still doesn't get it. The thrill comes through when he finally 'hears' you, and the broad smile breaks out on his face as he realizes 'I CAN DO IT.' The instructor can finally relax and safely send him solo.

A flying instructor training *ab initio* students must let the students make the usual mistakes, let them recognize the mistake and then let them take corrective action. To help a student build confidence in his ability, a flying instructor will appear to leave the student alone to do the exercise him/herself while trying to remain detached and fully relaxed. This stage happens, of course, after the student understands and demonstrates knowledge of the drill.

There is a fine line between recognizing that the student is flying into a dangerous position and letting him think his way out on his own. If he makes it, he will gain experience and confidence that will serve him well in future. But the instructor has to draw the fine line instinctively.

I made a serious 'fine line' error years later when I was a more experienced pilot, long after my flying school days. I was teaching our middle son, Jim, to fly in our old Taylorcraft BC 12D. CF-BMY was a tail dragger with two seats side by side, a big round control wheel and a 65 horsepower engine. Jim was nervous about his landings. A crosswind on the asphalt runway at Ottawa's Rockcliffe Airport was bothering him. To assure him that I had complete confidence in his ability, I kept well clear of the controls and had my legs crossed on final approach so he would know I had no intention of helping him keep straight after landing in the crosswind. A gust caught the aircraft just before touchdown and the Taylorcraft started to swing as soon as the wheels touched. I thought a ground loop was imminent. Quickly, I tried to uncross my legs. My knee caught on the control wheel. I kicked the rudder pedal to straighten the rollout as the aircraft headed off the runway. It was the wrong rudder pedal! The aircraft was still rolling about 25 miles an hour when the full impact of rudder and brake hit on the wrong side. The airplane ground-looped, turning a full 180 degrees. We completed the landing run backwards. In a heavier or a more modern aircraft it would have torn off the undercarriage, broken the propeller and dragged one wing tip. Not so the rugged T-craft.

I suffered much humiliation. My son, whose confidence I was trying to build, was terrified, but the aircraft suffered not at all. This stupidity on my part was a real downer for me, as I had undone what I was trying to achieve. Giving the appearance of being relaxed is important in teaching flying, but putting yourself in a position where you are not able to help if needed is stupid and unfair to your student.

Many lesser men would have quit flying after such a scary manoeuvre so early in their flying career, but Jim went on to become an excellent pilot who enjoys private flying immensely. However, he did not decide to make it a career - at which he would have been very successful. He has taught his sons to enjoy airplanes, so the skills will be handed on to the next generation. Jim is an Air Traffic Controller and periodically writes articles on aviation for publication.

It is easy to 'lose' students during the early stages of flight training if the instructor is insensitive to the students' feelings. Many times young passengers are taken for local flights by well meaning but macho pilots who try to impress their passengers with their skill and bravery. Most of these passengers become afraid and do not come back again. Such show-off flights often instill a lifetime fear of flying. Regrettably, whether because of a poor instructor or some other reason, they have missed the joy of flying.

Today, many of the better flying schools are associated with a college or university. Students take academic training along with the flight training and graduate with a degree plus a commercial pilot's licence, and (if their parents can still afford it) an instrument rating and licence to fly multi-engine aircraft, both prerequisites for consideration for a job with any airline. Some bush pilots fly their whole career without formal instrument training. The problem with a training program, for independent pilots, is that it costs about $1,000 to renew an instrument rating annually after completing the initial course. The cost goes for the recurrent training and renting an aircraft for the test ride with a government inspector. Established companies have an internal training program, the efficiency of which depends on the training budget and the degree of professionalism of the chief pilot and the training department.

In retrospect, this period of instructing was of great value in my career. As you never master an academic subject thoroughly until you teach it to someone else, instructing tends to improve your flying standards immensely. All your students are watching you. All your demonstrations must be correct, precise and accurate, or you lose credibility. In the same way, when teaching aviation ground school you must do your own homework well in advance or you will be asked embarrassing questions by the brighter students.

Not long after the war, Canada was a signatory to the International Convention of Air Navigation, a body of the United Nations dealing with standardization of all aviation procedures, airways, licensing, airports, etc. To meet the new standards, we all had to rewrite our licences, which was easy for me as a current instructor teaching just such theories daily - in two languages! Some pre-war bush pilots that had gone straight into the air force as instructors had never had proper ground training on subjects such as theory of flight, airframes, engines, meteorology and instrument training. There were some gaps in their theoretical knowledge through no fault of their own, and they were bitter after having done a superb job training the young aircrews during the war to be asked to re-apply for the new licences.

I had no problems and obtained licence number YZA 719. 'YZ' is the Toronto office of DOT designation; 'A' means Airline Transport, the highest licence available; and the number 719 means there were just over seven hundred who had this licence before mine was issued. A similar licence issued today would have six or seven digits. My engineer's licence also had to be updated and I was issued number ULM 158 - UL from the Montreal office, and the M means Aircraft Maintenance Engineer's licence, replacing the old A and C licences. These ICAO category licences were to permit us to qualify for a licence in any country subscribing to ICAO or about 95 per cent of the world community by simply writing the Air Regulations examination of that state. This was a great asset in later years, operating in various countries when on aerial survey and mapping contracts overseas.

We were very poor during the flying school period, but I had learned a lot that was to serve me well in later years. Although we went broke eventually, we paid all our bills. This part of my career is probably the most meaningful, in retrospect. As a man with a desire to teach and guide young people in some productive direction, this flying school was a high point in my life. The school set a high standard of training and we launched many young commercial pilots and instructors on productive careers in aviation. Many of them are now retired airline captains. This era also gave our three sons groundwork in aviation, which they nurtured and are passing on to a new generation.

In retrospect, I was probably enjoying the flying and family life at home more than I was paying attention to business. Suddenly, one fall, all the engines seemed to need an overhaul at once. Business had dropped off and groceries suddenly seemed very expensive. A career change was mandatory.

ALAN MACNUTT

The Canadair Days

My lucky introduction to the jet age

We were not at all happy to leave the moribund Champlain Air Services. It had been a wonderful experience but not a financial bonanza. It was time to approach a potential employer in the aviation business, one that would provide meaningful work and a meaningful pay cheque. And so, in the next leg of our westward trek, I stopped in at Canadair Limited in Montreal. My timing could not have been better.

In the early 1950s, Canadair employed thousands of people producing top quality jets for the military. During this period, the company was owned by General Dynamics, whose CEO, Floyd Odlum, was physically handicapped. His lack of mobility obliged Canadair executives to make frequent trips to the New York, Wilmington and Washington areas for briefings and to lobby politicians. The company executive aircraft was the time-honoured DC3 - CF-DXU - which had a crew of three, two co-captains and a flight engineer. The flight engineer's job was basically as babysitter for the DC3, 24 hours a day. He was to ensure that the aircraft was always clean, serviceable, the bar stocked and available to leave for anywhere, any time. The flight engineer needed to have an aircraft engineer's licence to supervise the aircraft's maintenance. He was required to have a clean shirt and clean fingernails when a trip was called. He had to act as steward, baggage handler and spare co-pilot. He had to be able to carry bodies, as any executive pilot will understand. The plane carried businessmen and politicians who talked freely after an abundance of gin and reduced oxygen. (Pressurized cabins had not been invented when the DC3 was designed.) To put it more delicately, some of our passengers had difficulty walking after long trips and often required assistance — all part of the service.

Now to my excellent timing. An hour before I walked into Flight Operations that the morning, they had just discovered that the flight engineer did not have a valid maintenance licence for the DC3. His periodic inspections had been illegal for the past year, jeopardizing their senior brass and posing a potential insurance hazard in case of an accident. He had just been fired. I'm sure there were more qualified applicants on their files, but the aircraft was due to leave for Wilmington, Delaware, in three hours. Since I had all the basic qualifications, the job was mine — if I could leave in three hours. No problem. The salary sounded immense after years of semi-starvation running my own flying school. Shortly after this stroke of luck we sold our aircraft and moved to Montreal. We regained solvency and enjoyed life in Canada's most exciting city.

The job was fun. I met a lot of interesting people that I would otherwise have only read about or heard about in the news. The job paid well but provided very little flying. As a result of my limited air time on the DC3, I got permission to start a flying club which we called CERA (Canadair Employees Recreational Association). It got off the ground with the purchase of an old Piper. The following year I moved to the Flight Test section. That was a far more exciting job. I got a lot of experience riding shotgun on over 300 initial test flights on T33s. Aerobatics was very much a part of our test flights, as we took each aircraft through the entire envelope of manoeuvres and tests to ensure it was in perfect shape at time of acceptance by the RCAF crew.

The T33 was named the Silver Star by our Air Force but never called that by its users who dubbed it the 'T-Bird'. This offshoot of the American P80 fighter jet was developed by Canadair as a two-seat (tandem) single-engine jet trainer for the RCAF. Eventually it was succeeded by the Canadair-built Tutor, a side-by-side jet trainer. The T33 was a beautiful little aircraft to fly -- manoeuvrable, rugged and demanding relatively low maintenance. We cranked them out at a rate of one a day during Canadair's peak. F86s and other models rolled out at a lesser rate. But I never got to fly the fabulous F86 Sabre that our test pilots loved so much.

The Canadair F86 was the American version in which Canadair installed the Canadian-built Orenda engine that gave it much better performance than the original. It was a single seat jet fighter, with swept-back wings that could achieve Mach One -- the speed of sound.

My new job involved flying on test flights of all new T33s. There had been problems in the RCAF after some of the trainers were delivered. No one had ever flown in the rear seat to ensure that instruments and controls functioned properly. That provided me the great opportunity to join the jet set, albeit only riding shotgun. The test pilots were great guys to fly with and I got excellent jet training. Some of them let me do the ride from the front seat. Not only was it a well paid, fun job, but also it gave me an entry into state-of-the-art technology.

It seems ironic that in his biography, Chuck Yeager does not give Canadair credit for the assistance given to him and Jackie Cochran in their feats in being the first in their respective genders to break the sound barrier in the Sabre. Both used a Canadair built F86 with the big Orenda engine. Bill Longhurst, our chief test pilot, spent time with them both, but Yeager does not give him any credit for this training. Yeager's Mach One was a major event in the fifties but is commonplace today. Al Lilly, a Canadair test pilot and later vice-president, was credited with being the first Canadian to break the barrier in a Sabre which many assumed at that time to be an impenetrable obstacle. The British had given up trying for supersonic speeds after Geoffrey de Havilland was killed at only Mach .94, but came back later with the Concorde, a winner politically if not financially. Mach is the speed of sound (1,100 ft/sec. or 750 mph) which varies with temperature and altitude.

At this time, the Canadair's Noorduyn plant (named for Norseman designer Bob Noorduyn) had just closed. The company had stopped converting Cansos from military to civil configuration and had completed modifications on the RCAF Dakotas to DC3 passenger liners or C47s as freighters. In addition to the T33 line and the F86 line, there was a booming spare parts business as well as surveillance drone manufacturing. Much interest was focussed on tests on the VT0L CL84, another Canadian aviation breakthrough that went down

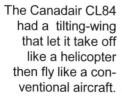
The Canadair CL84 had a tilting-wing that let it take off like a helicopter then fly like a conventional aircraft.

to political defeat. One oddball aircraft on the line was the C5, a Canadair mixture of the Douglas DC4 and DC6 with parts from each. The RCAF used it as a VIP transport but it was one of a kind. I made a few trips on it as flight engineer assistant to Gerry LeGrave.

Bill Longhurst was our chief test pilot. A rugged bush pilot from the Laurentian area, he later studied to become a doctor. 'Bud' Scouten, our second in command, came from an airline in South America and later retired to grow peaches in the Okanagan Valley of British Columbia. Ian McTavish, a former KLM pilot, died at an early age. Glen Dynes was killed test flying an F101 Voodoo later, in the Toronto area. I especially liked flying with Glen because he would occasionally

Clean fingernails was one of the tough demands for Al's job as mechanic, bar steward and co-pilot at Canadair. Helping stability-challenged VIPs was a small price to pay for the windfall job.

let me fly from the front seat, particularly the mornings after he had enjoyed a party. I savoured the pilot-in-command rush even if it was illegal.

Much of our flying was over the Laurentian Mountains that form a scenic backdrop for Montreal and a recreational area North of the St. Lawrence River. Advertised as 'the largest concentration of holiday resorts in North America', *les Laurentides* is indeed a beautiful tourist area of lakes, mountains, romantic *auberges* and modern hotels that provide a year round supply of great food, great entertainment and charming people. Forget politics; this area is happy land. You may come away broke, but you will have enjoyed every minute, and probably have gained a few pounds.

One of the places in this area that is most enjoyable, especially to old bush pilots, is Lac Ouimet near St. Jovite. This was the home of Tom Wheeler, one of Canada's legendary bush pilots and the locale of the airline and a couple of hotels he founded. Gray Rocks Inn may be getting a bit seedy with age, but it carries much history for the bush pilots of the 1940s and 1950s. Gray Rocks is near Mont Tremblant, the famous ski area just an hour's drive from Montreal. Thrills on the hills are common, but rarely do skiers get to see 'test flights' up close.

One day, a friend of Glen's had gone out skiing on Mont Tremblant, only a few minutes by jet from Montreal. Glen decided to impress her. He flew around the back of the mountain and up the blind side, rolled the airplane onto its back at the exact top of the hill, and flew down the ski slope side upside down about 50 feet above the skiers and the people on the ski lift. As we reached the bottom of the slope he rolled out, at low level and low speed, and flew away. Imagine going up a ski hill on the lift and looking up and seeing a sleek jet aircraft a few feet above you, going the opposite way down the hill, with two guys grinning at you.

Naturally, he did not go back and do it again. When a pilot does something silly at low level to impress people on the ground, he never goes back. The first time, everyone is gaping in disbelief. If you go back, some sorehead will take your registration and you'll be on the mat for all sorts of violations of air regulations. I have never seen this inverted downhill stunt before or since, and I don't recommend it for amateurs.

Glen told me later that his girlfriend, and the group with her, was suitably impressed. I know I was. It was a terrifying experience looking into the faces on the lift going the other way. But, like many flying experiences, the rush of excitement and high adrenaline flow provided a thrill I'll remember for a long, long time.

One of the test pilots with whom I flew in T33 testing was Hedley Everard, a veteran who maintained air force connections as a member of the RCAF Auxiliary. Since I had some jet experience and a fighter background, he invited me to join 401 squadron in St. Hubert just south of Montreal. Soon I was flying de Havilland Vampires, my first experience as pilot in command on jets. This meant an hour or two of pleasure flying a high-performance, jet, and playing war games on weekends. Along with this activity came the corresponding celebrations in the mess after our weekly flights.

My experiences with Canadair were all happy ones. I made a good salary so I could afford to pay my flying school debts. My wife had produced another handsome young potential pilot whom we called Jack. I had now acquired experience in aircraft manufacturing, executive transport flying, jet flight testing, and flown the T33 and the Vampire. Dave Fairbanks, later of de Havilland, Tommy Dowbiggan of CAE Industries, and Hedley Everard of Canadair, were our flight commanders in 401 Fighter Squadron. In Canadair, Smokey Harris, Johnny Munroe, Herb Gregoire, were all helpful mentors and gave me support and good advice when I needed it. It was one job I hated to leave. There was so much there to learn and through the CERA flying club I even had a chance to pass on some of the joys of flying to others.

But few pilots are happy to be second stringers for a prolonged period. It was time to get out of the government-subsidized environment and into the harsh realities of commercial aviation, where there is no one to cover your mistakes. Excuses don't count; you win or lose on your own merit.

Canadair had been a great boost financially. We were now debt free and had a few bucks for the next move up the ladder. The RCAF Auxiliary had been great fun and games, but the peacetime RCAF seemed like kids' games. Wartime flying manoeuvres without a war seemed a little like Alice in Wonderland. I needed some more real

Riding shotgun in the back seat of T33s gave Al an entry into the pure jet age.

flying and applied to Spartan Air Services in Ottawa, for a pilot's job in their rapidly growing aerial survey operation. I told my problems to Scotty McLean, a test pilot who had joined Canadair after an airline and executive flying career. Scotty called Weldy Phipps who had been

Hedley Everard (left) introduced Al to jet flying Vampires on 401 Auxiliary Squadron. The Vampire was his first jet pilot-in-command experience.

a colleague with Angus Morrison in a small post-war flying school at Uplands, called Atlas Aviation. Weldy, then Operations Manager and a rising star at Spartan, invited me to Ottawa. He hired me and I started as an engineer on the floor with his promise that I'd get the first empty seat available to fly with the company. It worked out well and I got my Lockheed Ventura, Canso and Mosquito endorsements on my engineer's licence all before I got to fly the aircraft. This was the second big step up in my career since the flying school. I was to get much valuable multi-engine experience on high-performance aircraft. I would be recognized as a professional pilot and a valuable asset to the company - a real ego booster.

A Spartan Mosquito crew takes a break in Fort Nelson, B.C. From the left: Len McHale, Scotty Gibson and Ted Lloyd.

ALAN MACNUTT

Spartan of Ottawa

World leader in mapping

When I joined Spartan in the fall of 1954, I worked on the hangar floor for the winter as an engineer. This gave me some idea of the company fleet and the people who made the company work. We moved to Ottawa and shortly thereafter could afford to buy our first home (with a $500 down payment and a 4½ per cent mortgage). We settled in for a happy new life in the nation's capital, which our family still calls 'home'.

The first pilot seat open was on a float-equipped Fairchild Husky based in Norman Wells, NWT. Hartley Marsh had just vacated the position. I had a checkout on the type by Weldy Phipps, whom I greatly admired. He ran me through some docking practice on the speedy waters of the MacKenzie River during spring run-off. Flying on floats in the Arctic is not the most prestigious job in aviation, but I was back in the saddle again. Later, I was checked out on a variety of Spartan aircraft. It included a tandem-gear Piper Super Cub (dual bogie wheels), a Dragon Rapide on photo survey, a staggerwing Beech 17S, then the mighty Anson, and finally the P38, Mosquito, Canso, and refresher training on the DC3. It was an exciting company in the 1950s and 1960s, and I was proud to be involved.

Over the early years of aviation, it seems most pilots got their jobs as a result of a recommendation from someone else. Scotty McLean had done me a big favour in getting me the Spartan job. Few pilots seem to have been hired because they had an inflated resumé or talked a good story. Hours of flying experience were often exaggerated and, in most cases, a pilot's reputation had preceded him for better or worse. A pilot's logbook is not a reliable document when applying for a job. It does, however, give the chief pilot the names of the people the applicant has known and who can be called for references. Pilots are extremely critical of one another's techniques. A navigator once told me that when an airplane takes off, every navigator present checks his watch; every pilot present does a critique of how the take-off was botched. Maybe we are just jealous of our privileged position and don't wish to share it. But such monitoring can eventually establish the job-hunting pilot's unofficial credentials.

My recommendation to general aviation pilots looking for a job in slow times are simple: If you have done a good job with your last employer, have worked hard and not whined unnecessarily you will probably be offered jobs. Or you might hear of them from people you have helped, impressed or befriended over the years. If all else fails, go north and take your lumps until the situation improves. In my early days after the war, you were not considered a serious pilot until you had rolled your share of gas barrels at some remote airstrips in the north. This was the training ground for most of the early Canadian Pacific Airline pilots, and it helped build a great airline from humble bush beginnings.

Weldy Phipps was a clever engineer as well as a pilot. His practical innovations building efficient survey/mapping platforms from war surplus aircraft put Spartan Air Services on the map. He later earned the Trans-Canada McKee Trophy for his outstanding contributions to civil aviation. Although his contributions were many, particular recognition was given for his pioneering of the large, low pressure tires that enabled conventional light aircraft to land on otherwise hostile surfaces such as tundra, pack ice, snow, river beds, shale outcrops and gravel eskers. This extension of light transportation is credited with opening the High Arctic to practical exploration. The Geological Survey of Canada credits the invention with a 50-fold increase in territory examined in a single summer, with a commensurate saving in cost. Norm Avery eloquently captures the genius of Weldy Phipps in his biography *'Whiskey Whiskey Papa – Chronicling the Exciting Life and Times of a Pilot's Pilot'.*

During Spartan's busy years, many countries did not have the accurate maps both military advisers and developers urgently required. They needed knowledge of the nations' geography for highway and rail construction, water drainage dams, flood control and agriculture. Spartan was on the cutting edge of survey at a key time in history.

Aerial photos for mapping depended on a stable platform on which the camera could be mounted in order to take a series of overlapping pictures. These were normally 60 per cent forward overlap and 30 per cent side overlap on the adjoining line. This permitted the mapping of detail on the ground. Each photo frame, when paired with the adjacent photo and viewed in stereo plotters would produce accurate three-dimensional detail. Towns, rivers and lakes could be plotted. Vertical elevations for highway construction, mining, log piles, ski slopes and a myriad of other everyday uses in engineering and land development were captured this way.

While Spartan was developing the photo platform from the war surplus fighter and reconnaissance aircraft, a Swiss company called Wild was developing state-of-the-art aerial cameras. They had a six- inch focal length to take pictures and advanced stereo plotters to display contours for mapping. This would replace the awkward and troublesome wartime cameras and the hand held stereo plotters previously used to describe land contours.

Engineer Bill May works on a P38 at Fort Smith, NWT. Note the Phipps nose bubble that accommodated the navigator.

ALTIMETER RISING

Under Weldy's direction, Bill Law and his fellow engineers hacked the noses off military Lockheed P38s to provide for a two-man photo crew. They installed a plastic bubble to house the navigator/camera operator on the nose of the aircraft where the guns would have been. An audio intercom established communication between the operator and pilot by which the operator could give courses to steer, altitudes to hold and other operational requirements. There was, however, no physical or visual link. We could not see each other and it was necessary to ensure two-way voice communication was kept up at regular intervals to preclude the danger of anoxia, or oxygen starvation. As we operated at altitudes of 30,000 up to 35,000 feet, lack of oxygen was a constant worry. If a pilot ran out of oxygen, or if his oxygen lines were pinched or punctured, he would almost immediately become erratic. The navigator would soon know there was a problem, but he could do little about it. The airplane would soon go out of control. Theoretically, the pilot would wake up disoriented as the airplane spun or spiralled through about the 15,000-foot level. By that time the man in the nose would have had second thoughts about how to make a living. There were no autopilots like those on modern airplanes; the aircraft was hand flown hour after hour to precise limits of 200 feet vertical and two to three degrees in azimuth for photo accuracy. This is not always easy in the thin air of the tropopause.

If the camera operator lost his oxygen, which could happen very easily just by jamming the soft rubber line with his elbow or knee, the pilot might not know for some time, so it was essential that we kept up a line of inane chatter to ensure the other crew member was behaving normally. A change of voice pitch, an unexplained giggle or slurring of speech, told the pilot to reduce altitude by about four miles pretty damn quick. The life expectancy of a human at 35,000 feet is very limited without oxygen. You would normally become unconscious in about two minutes. Modern aircraft are pressurized, but wartime aircraft required the use of oxygen carried in tanks and fed to a facemask through a regulator.

My favourite crew member was Vince Kluke, a veteran RCAF navigator from Barry's Bay, Ontario. Vince and I flew many miles together and I'm sure I scared him many times, but he rarely complained. Vince's job was to operate the camera, making sure he set the interval speed accurately based on our speed over the ground, ensuring that the pictures would be properly overlapped and the drift angle of the camera set properly. He also had to steer me by verbal commands such as 'two degrees right' or 'one degree left,' to keep me on the imaginary line over the terrain. As he did not have a map to navigate from in many cases, he would sketch the area on a pad as we flew so that he would know where to pass over on the next line in the opposite direction. In lake country this is relatively easy, but over rain forests, deserts or prairies this requires great skill on the navigator's part.

When I was first checked out in the P38 in 1955, I had been flying Ansons — quite a speed change in itself — and although I had flown several high performance aircraft, including jets, I did not have a lot of twin-engine experience. The P38 not only impressed me; it impressed the chap selected to check me out. He sat piggy back, literally on my shoulders, as it was a single cockpit aircraft that had been redesigned to accommodate an observer. After two circuits on the longest runway at Uplands he declared me ready for operations. He seemed anxious just to get out. I couldn't blame him; he had no dual controls, but had a wife and family.

Vince Kluke, a navigator taught Al the finer points of flying the P38.

The P38 Lightning was a high-performance twin-engine, twin-boomed, twin-tailed, high-speed aircraft that was produced in large numbers in the U.S. during the war and ferried or shipped to Russia, Africa and Europe. It had exhaust superchargers that gave good high-level performance up to about 35,000 feet above sea level. It was stable, easy to fly, heated, and comfortable, but required a lot of maintenance. They caused us problems keeping them serviceable in the remote areas of the north.

Next morning, after my check out, Vince and I fired up a P38 at 0600 and headed for Churchill on a photo job. First fuel stop was Kapuskasing, Ontario, a busy airport in those days on the Trans-Canada Airlines route. Although the runway at Uplands was probably 5,000 feet long (and during training I had used it all), the strip at Kapuskasing was about 3,000 with wires high above the approach end as you crossed the road on landing. My instructor had warned me to 'boil it in' at high speed. Another pilot had stalled a P38 at low speed on final approach and killed himself a short time before. That was to be avoided as airplanes cost money.

Being young, naive and stupid, I 'boiled it in' on the Kapuskasing runway, staying a safe height above the wires. I got the aircraft onto the ground about halfway down the runway with lots of speed but very little cement remaining. We came to rest finally in the cabbage (a pilot's term for off the strip) -- well past the end of the runway -- but managed to taxi back to the terminal, a very shaken crew indeed. Can you imagine Vince's position in the plastic eggshell nose watching this novice P38 pilot barrelling down the runway, brakes squealing and running off into who knows what cow pasture at the end? Poor guy! A normal man would have quit, taken the bus home and reconsidered his career options. Not Vince! He took me to the bar in Kapuskasing -- flying scrubbed for the day -- and spent the day and evening telling me how to fly a P38. Mind you, he had no pilot training, but as the evening wore on, my knowledge of the airplane seemed to increase. He literally taught me all I needed to know about the aircraft: speeds, approach, profiles, power settings, oxygen drills, and fire and escape procedures (we wore parachutes), and many facts that were missed in the company check-out. Next day we went to Churchill.

It was probably about one year later, flying with Vince, that we got another scare when we were working a photo job out of Winisk, Ontario, a Mid-Canada Line radar base on Hudson Bay. To improve communication in the aircraft, we had rigged up a Rube Goldberg device like an outdoor clothesline. It had a small pulley on each end, a loop of cord and clothes pins on which we could send each other notes, a pencil, or whatever was needed, or just a joke to break the boredom. One pulley was beside my seat within my peripheral vision with the front pulley just ahead of him. One day I

Spartan P38 was an early key to aerial mapping.

The Mosquito followed in the stable of high-altitude survey aircraft.

sent him a note that he didn't acknowledge. I called on the intercom - no answer! I rolled the P38 on its back pulled the nose down, inverted to about a 45-degree angle and dived at a high rate of speed to get Vince back where he would have some air to breathe as he had obviously run out of oxygen. At about 10,000 feet, he recovered consciousness and came on the intercom with a string of profanity that sounded like beautiful music to me. He was giving me hell for being off line, off altitude and fouling up the whole damn photo contract. I flew back to base and landed. We realized we were two very fortunate people. How long he had been unconscious we never knew, but our intercom chatter increased after that day.

In 1954, Spartan operated P38s off the sand strip at Sawmill Bay, NWT, on the south end of Great Bear Lake, near the Arctic Circle. Sawmill was a base for other photo and transport services. The USAF had abandoned thousands of barrels of aviation fuel there when the war ended. The fuel smelled like swamp water. The octane rating had deteriorated but the price was right. Spartan had to move on after a short stay because sand ingestion was hurting the engines nearly as much as the gas. But all was not strife at Sawmill. Recreation consisted of going a few miles east to Gunbarrel Inlet and catching Arctic Char for the table. The fish were huge and delicious and easy to catch.

Like the converted Mosquito bombers that were to follow, the P38 accounted for several fatalities during its postwar life as a photo ship. Mike Todderon and his navigator were killed in a test flight out of Ottawa one winter. Another pilot stalled on final approach during training. Jack Tustin and his brother went down in a Mosquito. Jimmie Lago, the wartime pilot of Weldy Phipps's Halifax bomber along with navigator Gene Benoit were killed when they ran out of fuel on approach to Dawson City in a P38. As in wartime the high fatality rate was taken in stride; lots more airplanes and pilots were available.

In 1956, Spartan recruited Roger Tessier, a timid young man from Kapuskasing, as a novice camera operator. When his training had been completed by Joe Scott, our senior cameraman, I was asked to take him on a photo job on a Mosquito to Mont Joli. I felt my responsibilities very greatly in those days, and spent a lot of the trip instructing Roger on his duties and responsibilities. I was worried that he might mess up the photo job as there was quite a lot of money involved and we had a fussy customer. Next morning, I hurried into the lab to check the pictures to see how Tessier had made out. I was told 'The pictures were great, but the flying was a little rough; altitude was poor and corrections too sharp. Who was the pilot on that job?' Roger, later a senior helicopter pilot with Alberta Forestry, has had a lot of laughs over that one.

He got his revenge out of Timmins later that year when he and I with Vince Kluke were doing photo in a Mosquito at 35,000 feet. When we landed from a four-hour morning trip, Vince and I would have to add oil and replenish our oxygen supply and hand pump about 700 gallons of gas into the aircraft, taking turns on the hand pump. Roger would take the old rented crew car and go to town and get us hamburgers and milk. Any kind of soda pop was a bad idea, as it bubbled up at high altitude and upset your stomach. Roger would always seem to return at the time we had just finished pumping the gas and were lying down for a few minutes break before taking off on our afternoon trip. After a few days, Vince and I realized we were doing things wrong and then we took turns on the hamburger haul. After this strenuous exercise, we would then fire up the Mosquito and try to eat our hamburger in the climb, without leaving off the oxygen mask.

In prolonged high altitude flying without pressurization, we always used oxygen from the ground up, trying to store as much oxygen in the blood stream as possible, even on the ground. Hockey players now do much the same. The 'bends', caused by the 'bubbles in the bloodstream' causes swelling in the joints and is very painful. Some high level photo crew members, when working at altitudes of over 30,000 feet, suffered more than others, but the smokers seemed worse off. They also had the problem of not being able to smoke when wearing an oxygen mask. Fire and oxygen together will cause a major explosion. In some cases, crew members could not stand the extreme altitudes without pressure. They went to other lines of work, such as low-level geophysical survey, or joined an airline. Some escaped back into the peacetime RCAF, which we called an early retirement. In retrospect, it seems some of us survived on ego, gasoline and lots of oxygen. Speaking of ego, I recall one day while refuelling somewhere in Northern Ontario en route north. The gas man, to make conversation, asked me if the chap with me, Cal Marshall, was my co-pilot. I blurted out, 'Oh no, he's only a cook!' It shows two things: first pilots are very egotistical, and secondly putting down the man who is going to be your cook for the summer is not smart. Cal had worked with bush pilots for years and saw the humour, fortunately.

Laughs on northern operations came in various colours and at a different potential cost. A joust with an Austin Airways crew comes to mind. Austin was a respected air carrier operating in Ontario and north around Hudson Bay into the Arctic. They also conducted aerial survey over the years for Inco, the mining corporation, often using a magnetometer towed on a cable behind an Anson. When not being used, the 'bird' would be pulled in by winch and would nestle in a cradle under the belly of their old red Anson. This 'bird' was about five feet long and about a foot in diameter and looked like a bomb. When we ferried magnetometer aircraft overseas, the 'mag' was carried inside the aircraft and on many occasions we had difficulty explaining this to customs authorities in foreign countries.

One of Austin's pilots was John Croker, a pleasant British lad. While we were based at Timmins, a good-natured rivalry resulted between him and his crew and my Mosquito crew. Being wooden, the Anson had less steel and electrical interference for magnetometer work, but the airplane was slow and cold. Our airplane was also made of wood and rags and also British designed, therefore cold, but much faster. That gave our crew points in the macho department.

Coming home to base one evening, we spotted John's crew making their way in the same direction. We flew in high above the Anson, shut down the right engine on our Mosquito and dived by his left window with the right prop feathered at an overtaking speed which was probably about 250 miles faster than the Anson. We pulled up sharply in front of him, started up the other engine when we were out of sight and waited for him back in Timmins to needle his crew. Point one for the Mossie crew. We explained to John that next time we'd come up under him when we saw him in 'our' airspace and do a loop around him.

About a week later, we saw the Anson again, ferrying home to base. We contacted John on the radio, and told him we were coming in behind him and to hold on for the bump as we came up in front of him and he would have to fly through our slipstream. He replied "OK, just tell me when you're half a mile back". As we came up under and behind at a high closing speed, John dropped the magnetometer from its nest under the Anson in a free fall right into the flight path of our Mosquito. We dodged the magnetometer and the cable on his bird took the shock without losing the $3,000 mag. We had a good laugh later, but we could have all been killed by a bomb in peace time. If our respective bosses had known, we would have both been fired. The score was now tied, Christians 1 - Lions 1, so we quit the shenanigans before someone got hurt.

For more tales of Austin Airways, there is a handsome book written by Larry Milberry called 'Austin Airways'. It traces the company from modest roots to airline prominence. It is a record of the many, many good bush pilots that worked for Austin in the period from pre-war onward.

One memorable red-faced incident occurred in Mont Joli where two Spartan crews were operating P38s on a photo job about 200 miles north of the St. Lawrence River. The two aircraft had been out on a recce and finding cloud in the area, scrubbed the photography, and returned to Mont Joli. The crew landed and went downtown for breakfast. Spartan president Johnny Roberts flew down the river a little later in the day in an Anson in clear skies. He spotted the two P38s sitting idle and unattended in fine photo weather, and thought the crews were goofing off. He wasn't expected. He couldn't raise anyone on the radio and roared in to land in a fit of temper, ready to fire the two crews. In his agitation, he forgot to lower the gear on the Anson and when the crews came back out after breakfast, they were surprised to see a company Anson — gear up — blocking the runway. They were still more surprised to see the president on the verge of a heart attack, not knowing who to blame, or for what.

One of the pleasant experiences I have encountered many times over the years is returning southward from northern or Arctic operations and seeing, probably for the first time in months, the first tree. Then as we continued to fly south, more trees and a farm or a road and fences and later towns and people. We get bored in towns with too many people, too much traffic, but humans have social instincts. It's a thrill on contract completion, to set 'S' on the compass and head for 'momma's house', as the truck drivers like to say on their CBs. To go home to clean sheets, good meals and family life is a pleasure I have experienced more than most people. Old Wally Scott said it best: "Breathes there the man, with soul so dead, who never to himself hath said, This is my own, my native land!…"

Spartan prospered for many years, but with the advent of the Learjet, which hit the market in the 1960s and which could fly higher and faster and provide a smoother, more stable camera platform and a pressurized work area, the Mosquitoes quickly became history. CF-HML became a museum item in Vancouver but has disappeared from display; CF-HMS was being rebuilt in Calgary; CF-HMQ crashed by me in a fiery landing at Pelly Lake, is being rebuilt at Windsor, Ont., by members of the Canadian Aviation Historical Society. The rest were junked or sold overseas.

Johnny Roberts had flown Mustangs on an RCAF Pathfinder squadron. Russ Hall, co-founder, had been a navigator with an aerial reconnaissance squadron whose objective was to return with photographs of enemy installations. The two, together with some imaginative staff, built a high-tech survey company that the Canadian government took seriously. The RCAF withdrew their Lancaster survey crews from the Canadian Arctic and gave Spartan the mapping contract. Photo Survey Company, later to become Kenting Aviation, was also formed at this time and gave Spartan some competition. PSC's crews did not seem as aggressive as Spartan's and, although the company lasted longer, the glory days of aerial survey went to Spartan, as well as the majority of the big contracts both in Canada and overseas.

Several of Spartan's key personnel left to form their own companies and careers. Joe Kohut started a company called Capital Air Surveys in Killaloe, Ontario. Len "Black Bart" Bartley formed his own Pathfinder mapping company. Evan Jones went to Kenting, Harry Smith left to start an aircraft maintenance company called Personal Plane Services, and Chris Falkner left to form an aviation company in Edmonton. Jack Fleming bought a Cessna 195 (called a Shaky Jake because of the Jacobs radial engine) and started his own business.

Roberts and Hall, with the wise guidance of Jimmy Wells and the innovative Weldy Phipps, had started a growth industry from a modest beginning that became famous worldwide in both geophysical and mapping survey. I consider myself privileged to have been part of that organization.

ALAN MACNUTT

The Flying Husky

Bush plane – no bush

In the mid 1950s, Canada's Arctic had few airstrips and navigation aids were very rare. The bush pilot was in his glory days – even without the bush. Then came the DEW Line, with a string of airports across the entire High Arctic area. Each had its own low-frequency non-directional beacon, voice communication and the availability, by prior arrangement, of fuel. A row of airstrips across the Mid-Canada Line came next with the corresponding aviation facilities at a more southerly latitude, about the 55th parallel. Finally, the mining companies put in their own aerodromes and beacons. Native settlements got community airports, major Indian bands got their strips, and all accommodated modern aircraft. Aklavik moved east to higher ground and became Inuvik, and eventually roads crept north, all spelling the downgrading of the bush pilot's role in Canada. By the 1970s you could count 10 or 12 turbine-powered Twin Otters in Inuvik or Resolute or Frobisher to meet the evening arrival of the CPAir Boeing 737.

The Twin Otters would carry in ore samples, homebound drill crews, mail, Eskimo soap stone carvings, and countless government inspectors from outlying areas. An hour later they would return to their bases on the tundra with mail, spare parts, roughnecks back from their pleasures in the Big Smoke, and more government inspectors.

Spartan was a significant Arctic player in those days. Each contract would involve different geography, different aircraft, different crews, and often an innovative approach to bring it all together. At one point,

ALTIMETER RISING

our company used Shoran stations in the Arctic to provide control for the aerial photography. Two Shoran stations, each transmitting a signal from known positions, were received by the aircraft. Then, by means of triangulation we could fly a precise line and produce photography that could be turned into an accurate mosaic. The photos taken from a Ventura were both topographically and geodetically controlled in this manner. Shoran, as a means of control, was later updated by various systems, such as Doppler, Decca, and Inertial Navigation as technology evolved. The later Global Positioning System (GPS) would have done a better job for a fraction of the cost and maintenance.

When Learjets started doing photography, the old piston aircraft became obsolete. Eventually, satellites took over. In 1994, Canada's Radarsat was launched into space 500 miles above the earth, providing another leap forward in world mapping and surveillance. Radar can penetrate fog, and so the old problem of waiting for a cloud-free day for photography had been eliminated. All this meant the end of photo survey companies as we knew them. Now, Radarsat has been updated.

Starting in 1954, during the period when the DEW Line was being built, we had a Shoran contract in the Arctic. As the new boy on the block I was sent, with a float plane, from my base in Norman Wells on the MacKenzie River to Coppermine on Coronation Gulf to pick up a Shoran camp and move it east to Bathurst Inlet. The aircraft, a Fairchild Husky, was a fine old float plane built in Lonqueuil, Que. for the bush market, with intent to replace the aging Norseman. de Havilland had built the Beaver at about the same time and the two aircraft were competitive. The U.S. Air Force placed an order for Beavers, which assured their success and the old Husky died, so to speak. The Husky was powered by a Pratt & Whitney Wasp Junior that provided 450 hp for take-off. Most people considered the airplane under powered but that was not really true; it was simply over-fuselaged. The cabin was enormous and, of course, the tendency was always to fill it. If you had miles of take-off surface and didn't mind abusing the engine, the Husky would carry a very large load for the power. As a float plane it was normally loaded until it would start to sink. When the floats started to go under water at the dock we knew it was loaded, then we removed a few hundred pounds and tried to get it airborne, and tried . . . and tried . . . Later models were equipped with the Alvis Leonadis engine.

Clyde Blythe, a veteran bush pilot from B.C., flew the Husky with the British engine. He reported that the extra 100 horsepower and three-blade prop increased the Husky's performance a great deal. The new engine, however, required intensive maintenance.

The Husky's voluminous cargo space made it easy to overload. It wasn't under-powered, it was just over-fuselaged

Canadian Bushplane Heritage Centre

In 1954, I was stick handling a Husky (CF-BQC) under a low overcast from Yellowknife to our camp on Bathurst Inlet, a distance of about 600 miles. It was late summer, the light poor, and the terrain uninviting. Above the tree line the features vary little in colour. In most areas there are few lakes or major rivers and no major hills by which to navigate. As the ground is generally flat in the coastal Arctic, the lakes are shallow and may hide boulders the size of a Volkswagen. Only where the area is hilly can you be reasonably assured the lakes will be deep enough for the safe manoeuvring of a loaded float plane. The ground is usually marshy with poor drainage and permafrost only a few inches down, preventing normal water runoff. As the Arctic is basically a desert - very little precipitation in rain or snow - it changes little from year-to-year. On the surface there are frequently lumps of topsoil with clumps of vegetation about a foot across, which makes walking difficult. The terrain is impossible to land on with wheels. Occasionally a river bed or an esker can be found to land a light plane if it is equipped with the big Phipps Special tires.

The Husky I was flying had many characteristics that endeared it to its pilots. One was a theoretical seven-hour fuel supply. If filled, however, the weight precluded taking any meaningful payload. But the range was a godsend on a long ferry flight with so few landing areas. In addition, it was one of very few aircraft that you could refuel in flight. There was a filler cap in the rear cabin that let you top off the rear tank in flight from barrels in the cabin. This all assumed you (a) had a non-smoking helper, (b) had a funnel, and (c) it was not too rough. Turbulence would dump fuel all over the cabin - a scary thought.

The rear-loading Husky gave it one advantage over other aircraft.

As I made my way alone towards Bathurst Inlet the light was fading and there were no safe landing areas in sight. If a lake did go by it was gone before I was able to determine if it was long enough and had no hidden granite lumps to tear my floats off. To try and turn back at low altitude in bad visibility to find the lake would be very hazardous and you might not find it again in any case. I had lots of fuel spread over two additional tanks, but I had been using the same one for a long time. I knew that the lowering ceiling could turn to fog at any time, and that diverted my attention from my fuel condition.

With about 50 feet of air below me, I had my eyeballs against the windscreen trying to penetrate the clag when the trusty Wasp Jr. sputtered and quit. I had failed to notice the gauge settling on the Big E, the fuel pressure gauge dropping off, and the warning light go red. The pilots' joke of hours of boredom interspersed with seconds of pure panic applied. United States Civil Aviation researchers have found that the average pilot takes at least three seconds to react in an emergency. During that time he registers shock, disbelief, confusion and panic and 'Why me?' before he finally starts to correct the problem. In this case the three seconds seemed much longer.

I switched tanks and grabbed the hand wobble pump. Its job was to build up pressure to start the engine and supply fuel if the engine pump failed in flight. Unfortunately, the pump takes several seconds to pick up the supply. I got pumping and tried to keep raising the nose at the same time as the speed decreased. The unscheduled landing that was such a clear possibility would have meant a certain cartwheel into early retirement.

To keep my health intact I kept pulling back on the wheel to keep the floats from touching the tundra. This, of course, reduced the speed dramatically as the prop was in 'drag', not power mode and a stall was imminent. Finally, the trusty engine started to roar again when I was only feet off the ground and gradually I got back into a reasonable flying altitude with full power. With my more pressing chores, I had had no time to throttle back before the power came on again with the renewed fuel supply. This may have been a good thing as this was no time for reduced power and careful by-the-book engine handling. The next lake I saw didn't look great but I cut power, straight in, full stop, out anchor, and glad to be there. What rocks? That's where I spent the night in my comfy sleeping bag.

The moral is don't let a preoccupation with one problem snowball you into a bigger problem. Who knows how many accidents could have been avoided over the years if pilots could remember the big picture? When we have an immediate problem demanding our constant attention, the preoccupation often sets us up for a much greater problem.

ALTIMETER RISING

In those days, particularly on northern operations, you developed a closeness with your aircraft that is unknown to military or airline pilots. When you are flying for an airline, you walk off at the end of your flight and another crew is waiting to climb on board and fly it away. You may not see that particular aircraft again for weeks or months. In the military, if your aircraft is sick, you simply take another one. In bush flying, that aircraft is yours. It becomes part of your personality. If it goes sick, you are unemployed. If you damage it, you are fired! This feeling of intimacy is one of the pleasures and when the aircraft does well on a difficult assignment, you share the pride and the glory. The pleasure of punching up through the clouds into the beautiful sunlight dancing on the cloud tops is a shared enjoyment, and landing on smooth water or snow so you can barely hear the floats or skis, is a shared pleasure. Sometimes the aircraft seems to take as much pride in a good performance as the pilot does.

One day I was loading gas barrels in my Husky, rolling them from shore up a plank ramp into the rear fuselage loading door. A barrel got away from me and about 400 lbs. came down on my foot. My foot was soon badly swollen. Naturally, the thought of a trip home to Ottawa to see my beautiful blonde came to mind, but that was not to be. One of our P38 navigators at that base was 'Doc' Pocheriva, a medical student who worked for Spartan each summer as a photo navigator.

Doc bandaged me up and I spent a couple of days in Norman Wells bugging our cook, Cal Marshall, while Alf Lord was assigned to fly my Husky. Alf is a kind, gentle man, who was then flying a Ventura on photo triangulation, but was available while his aircraft was out of service for a few days. Although he was an excellent pilot, very thorough and careful, his experience had been on heavier aircraft with a navigator on board. Map reading and dead reckoning navigation over long distances, at low level with poor maps, was not his forte. In two days he was lost in the barrens. He could tell base camp by radio that he was safely down on a lake, but didn't know where. By the next morning his battery was dead, communications ceased and the search began.

One of the key items in safety gear for our crews was the Gibson Girl, a radio transmitter in a curvaceous shape that you held in your lap. Power was provided by a hand crank to send out an emergency transmission on 500 KC. Any search aircraft using a conventional automatic direction finder could monitor the distress signal. Alf cranked diligently and, with our entire photo crews looking, we found him on day three sitting on a lake in the Husky. He was badly in need of mosquito repellent, food and aviation fuel, in that order.

My foot was sore, but my eyes were still good in those days. I was in the right seat of a DC3 with Mac MacIntosh as captain, when we picked up Alf's transmission. Two hours later we had homed in on him, flew over the lake and marked the spot. A Norseman was sent in to pick him up, badly fly bitten but happy. I believe Willie Lascerich was flying the Norseman, but it could have been Brock Parsons. Next day I was back on my Husky again, sore foot and all, but I had a helper to assist with the gas barrels.

It seemed simple in those days to find a lost aircraft. Everyone in that search crew was personally involved. We were all looking for a close friend and operating in an environment we understood. Motivation is everything in such a search. We looked after each other but we also had some empathy with other residents of the north who relied so heavily on air transport, their lifeline to the outside world.

Once I was in Churchill getting a load of fresh supplies for our Shoran camps. When I was leaving to go north, I picked up the mail for points en route as was customary. In remote areas, people get mail any way they can. When I got over Baker Lake, a strong wind was blowing from the south and the waves near shore were too high for a safe landing. There is a small lake north of the camp, but I was in a hurry to beat darkness into Pelly Lake and decided to drop the mail bags on the beach in front of the Hudson Bay Post. Naturally, the whole town turns out for the arrival of an airplane — any airplane — but especially an airplane with their mail. I called the radio operator and passed the word along that I would fly low along the beach and kick out their mail bag in flight.

At the precise time of 'Bombs Away' I kicked the bag out of the door. But instead of my clever solution to the problem it got worse. The cord on the mail bag caught on my water rudder cable and wouldn't let go. I turned out over the lake and climbed into the wind to figure out what to do. But fate took a hand. In the turn, well out over the water, the bag released and plummeted into to the ocean and sank. The entire village was unhappy. If the wind had suddenly died, it would not have been safe for me to land. My local fan club was in revolt.

The bag probably contained lots of Eaton's catalogues to increase the weight. Regular carriers got paid by the pound for carrying mail, so they ensured everyone was on the catalogue mailing list. The cargo was rescued, with no damage, by some brave and very wet Eskimos and all ended well. The Anglican priest's new glasses were in the bag and did not break in the fall. He would, no doubt, have had me on his hit list if they had. On the other hand, the Catholic priest, who lived a few yards from him in the tiny community, would have been delighted. The two were bitter enemies and poor examples of the 'love thy neighbour' concept. Sandy Noonan was the Hudson's Bay factor in Baker Lake at that time. I can't recall the name of the only other white resident, the RCMP constable.

In the fall of 1954 I was ferrying the Husky back to Ottawa from Yellowknife and overtook a cold front moving from the Churchill area into Northern Ontario. The weather was terrible, a ceiling of 200 to 300 feet and very poor forward visibility. That part of the country, once known as Rupertsland is very flat and barren and the many rivers draining the area twist over the flat terrain finding their way to Hudson Bay. It is boring country to fly over with no navigation aids. Every river looks the same. The advantage under such conditions is the fact you are sure not to hit a hill or a tree and certainly not a television tower.

We used a trailing antenna to maintain HF communications and faced the danger when flying beneath low cloud of losing the antenna by snagging it on the ground. We would then be unable to advise base or a DOT radio facility of a change in the flight note. A flight note is a poor man's flight plan. So if you lost your antenna and had to land away from a base or town to wait out weather you might have Search and Rescue looking for you. Along the south shore of Hudson Bay there were few towns, fewer pay phones and no established float bases.

I went back north to follow the coastline to avoid getting lost over the featureless terrain. I made my way slowly along the shore eastward to James Bay and turned south heading for the Austin Airways base at Moosonee. This kind of flying is not pleasant. You are sitting forward in your seat, eyes out front, looking for islands or rocks, moving slowly because of reduced power and partial flaps, always hoping it will clear soon and you'll be out of the weather. High humidity can cause lots of carburettor icing and condensation on the windshield adding to the visibility woes. Earlier in the day, before I encountered the low cloud, great flocks of Canada geese were moving south, darkening the sky and adding to the reduced visibility. As the weather deteriorated they seemed to have used their natural good sense and landed for the day.

After about a five hour trip, I finally stickhandled into the river at Moosonee and tied up at the Austin Airways wharf, tired and glad to be able to relax. Almost immediately a rather large group of people crowded around the dock asking questions: 'Where did you come from?' 'Where are you going?' 'How was the weather?' 'Are you available for charters?' I was somewhat unhorsed by this sudden and enthusiastic barrage of questions. But it soon became clear. Every year during the goose season, large numbers of affluent hunters take a vacation and go out to the goose camps around James Bay to shoot their quota of geese. These people at the dock proved to be a group of these businessmen from Toronto and New York State. They had been waiting three days to fly out to the camps but had been grounded by the bad weather. They had flown into Timmins, taken the train from Cochrane to Moosonee and were annoyed to be wasting this time at a seaplane base with the bush pilots telling them the weather was unsafe. 'After all, the guy that just landed in the Husky is a bush pilot…'

It took some time to convince these businessmen that indeed the weather was too bad to fly and that I had only got in by good luck and they were better off on the ground. The local bush pilots relaxed a bit. The pressure was off as I had caused them some embarrassment by merely turning up alive. They knew this country well, which I did not. By arriving unexpectedly I had made them look bad in front of their customers. But my description of the weather had let the local pilots off the hook and they could now go away saying 'That silly bugger will kill himself one day!'

The pilots' fraternity, unlike doctors, is usually the harshest critic of other pilots. Many pilots will criticize another pilot's landings, take-offs and general flying at great length in front of passengers, but rarely in front of the pilot himself. This may be based on our own insecurity or the tendency to think of ourselves as the best pilot in the world, or some such nonsense. The pilot that says he never makes mistakes is a very big liar; the secret is to make small ones and live to keep on making them.

The joy of flying will sometimes push us into ridiculous situations. Fortunately, I have usually had the final decision in my hands whether to go or not go if the weather was bad, or an engine was sick, or the load was over limits. Many young pilots are not so fortunate. An over-eager dispatcher can often send a young pilot into situations he can't handle, by simply saying 'Well Joe had no problem, why can't you go?' or more often 'If you won't take it, we'll hire someone who will.' Strangely, many such companies have enjoyed remarkable success, have evaded union organizers, and had little trouble hiring pilots when required.

My load out of Coppermine one day was to be the Shoran transmitter and mast, the auxiliary power unit and the station operator. He usually was a bearded, wild-eyed university student on summer employment and suffering from isolation and extreme mosquito bites. Additional cargo included generator fuel, food, tent, axes, tools, boxes of radio tubes, antennae, guy wires, VHF radio, and many more items. The student usually had a hundred pounds of books for his course which he studied to help avert cabin fever.

When the station had served its purpose, the operator was advised by radio to break camp and carry all this gear down to the water's edge. This was about two days' very hard work, all by backpack. The Husky would then be dispatched to pick up the load and move it to the next camp where the operator would have to carry it all back up to another hilltop, set it up and get back on the air so operations could continue. Losing one small item such as a box of tubes, or a battery, could render the station useless and could bring the whole project to a halt. The Arctic summer is fleeting and a down day was bad news for the company. Nowadays, a similar move would take place by helicopter, and station down time would be minimal. Replacement parts came from Ottawa, and as there was no airline service to the area, parts replacement was an expensive and time consuming business. Anyone losing a key component was not especially loved. But we had one special way to 'lose' components. It was a method that I am ashamed to relate now in this 'green' generation, was to seat one of my passengers near the rear door. The plane had four doors plus a huge hatch in the rear for loading gas barrels and canoes. He was briefed to start throwing payload out of the door during the take-off run on pre-determined signals, if it appeared we were not going to get airborne, or, worse still, if we were showing signs of sinking.

We were not in the least pollution conscious. It was not unusual to throw out 10-gallon barrels of naphtha gas, maybe a blowpot or two, a tent or case or two of canned fruit or spam to facilitate our take off. The only reason we hesitated to do this was that we might have to abort the take-off and taxi back for another attempt. On the second run in Arctic twilight we might hit one of these objects. Hitting a blowpot or a can of naphtha would, of course, puncture a float and we would sink very rapidly. The items we jettisoned would eventually float up on shore and would be a real prize for Eskimos when found months or years later.

The results of such littering are now being felt in the Arctic. What was once pristine barren land is now littered with empty, rusting gasoline barrels, discarded snowmobiles, jeeps and assorted empty tin cans. A blanket of smog from factories in Europe and Asia is now reported over large areas of the High Arctic and seals are said to have high dioxin levels. On the brighter side, millions of dollars have been spent to clean up the Distant Early Warning sites and other bases used by the U.S. Army Air Force.

During this particular contract, the old Venturas were used as photo ships. People such as Alf Lord, Jerry Bruce, Ken Fraser and Sam

Taylor, chief pilot, flew them. People like Tom Patterson and 'Black Mac' MacIntosh would fly in the gasoline in 45-gallon barrels in a DC3. Grant Mervyn and Donald Davidson would keep the electronics serviceable. Jim Murray, later of Ducey Aviation in Edmonton, would make sure the radios and the vast array of electronic gear was in working order. Gordie Townsend, Neil Armstrong and John Theillman would later lend helicopter support.

Cal Marshall and his wife Ann would run the cookhouse. Mechanics would keep the airplanes serviceable. Many people worked together, each in his own trade, to make mapping of the Arctic possible. Previously, a Lancaster squadron based in Rockliffe was doing Arctic photography, but it was replaced when Spartan and Kenting came on line in the 1950s.

Many man hours of work went into making the maps we now take for granted. Dr. George Zarzycki was our photogrammatist in those days in Spartan and his high standards of work back at our Ottawa Base in Uplands resulted in the accurate maps we now enjoy. However, geography changes constantly, and map making is an ongoing process. With satellites that can read a car licence plate from a distance of many miles, we get some idea of how this industry has evolved in recent years. Not long ago, most mapping was done by men on horseback who climbed to hill tops and sketched what they saw. Fifty years' progress in a high tech industry is a long, long time.

With vast distances without landing areas, and cruise speeds of little over 100 miles per hour make flights over the barrens long and lonely. I recall often listening on HF (High Frequency - long range) and occasionally hearing another lonely bush pilot trying to find someone to chat with to break the boredom. Brock Parsons, a veteran northern pilot who flew out of Yellowknife, had his own particular signature on the radio for reporting 'all's well'. Brock would simply open the mike and sing a few bars of "I'm a-rollin'..."

One of my pastimes was to read paperback books. I would hold the book where the sun shone on it at a certain angle. Any deviations in course or attitude of flight would immediately cause a change in the sun-shadow pattern on the page I was reading. When the page was completed, I would tear it out and put it beside me on the floor, do a cockpit scan of instruments, fuel, position, and back to my book - hour after hour.

Harry Smith, our newly-immigrated engineer from Britain, would always know I was flying when a float plane landed at the base at Norman Wells. The door opened and out came several chapters of my current paperback to be thrown into the river. Environment was probably not in our dictionary at that time.

Churchill — the Big Smoke

Not quite a 'plum' assignment

Churchill, Manitoba, was the supply base for Spartan's operations in Northern Canada for many years. The local whale factory was in full operation then and the smell of rotting beluga parts was part of the townsite's charm. Later, Edmonton emerged as a better centre; it had hangars, maintenance facilities and more than two hotels. The two firetrap hotels in Churchill were always full, noisy, and overheated. In the early days, civilian aircrews were welcome in the Officers Mess at the Churchill air force base, which meant good living conditions while in town. But when the RCAF became annoyed at some perceived misdemeanours of our crews, we would get thrown out of our mess status and have to sleep on the floor in sleeping bags with the rest of the crew, in whichever hotel would let us in that month. The floor of my aircraft was cleaner and more comfortable than the hotels in the summer, but the mosquitoes were terrible. In the winter that was not an option.

One summer Gordie Ewing was flying a Canso out of Churchill. Sam Taylor, Rocky Laroche and I were flying Mosquitoes and Jerry Bruce had a York there, all at the same time. It seems that Gordie had got there first and rented the only U-drive car available in town for his crew. It was a large luxury model Buick. The rest of us had to settle for wrecked pickups and ride sitting in the back exposed to the elements and the ubiquitous mosquitoes. He would never share his Buick with the rest of the crews, and that became a sore point in the group. One day I had a mag drop on one engine and our engineers asked me to do a run-up to check that it was fixed. At the same time Gordie drove up in his Buick. He parked just off my wing tip on the hard stand to avoid a gravel storm during a power run. When the prop tips got up to speed, a sonic ripple shattered the windshield in Gordie's Buick. Glass showered all over the seats, transforming the car into a wide-open fresh air convertible. Now you may be sure there was no Buick dealer in Churchill, nor was there a glass shop. For weeks, Gordie's crew drove about in their convertible, blaming me for their lowered status. My crew thought this was hilarious.

Some incidents were not so funny. A new pilot was hired who had far more experience than any of us. Jim Pattison, of Toronto, had been instructing Chinese Air Force pilots. The Chinese had purchased a large number of Mosquitoes, and Pattison had trained them before coming to us in Spartan. He knew a lot of good tricks about the aircraft and was willing to share them with us. He was a welcome addition to our company.

After an initiation to Spartan, he and Len Cooke, our chief pilot at the time, took off one evening in a Mosquito from Churchill to join our operation at Pelly Lake, a trip of about 750 miles. Cooke was to introduce him around, show him the Arctic and go back to Ottawa. They left Churchill in good weather and with a good forecast. During this post war period, I believe we had the best aviation weather forecasting we have ever had in Canada. The forecasters in Edmonton, Winnipeg, Churchill and Montreal in those days, were devoted to their task and took great pride in their work. They used all they had learned in wartime, all the information they could get from pilots, looked out the windows, and had a feel for a wind change or drop in spread that gave pilots great confidence.

When 'Cookie' and Pattison got to Pelly Lake the fog had rolled in off the lake. Although we had a low power beacon suitable for an instrument approach, they could not see the runway on final. After two missed approaches they decided to go back to Churchill while they still had fuel. The only alternative was Yellowknife. Although Yellowknife was closer, the crew did not have the maps or frequencies and they radioed us that they were en route Churchill. After getting landing clearance from Churchill tower, the Mossie ran out of fuel and the two pilots died in Hudson Bay. 'Doc' Desmerah found their remains later washed up on the shore, but only after a polar bear had found them first. We had many funerals in those days but with wartime conditions so fresh in everyone's mind, it did not seem so unusual as it would later when accidents were not considered an acceptable part of the aviation business.

It is regrettable that the crude FIDO (Fog Investigation Dispersal Operation) system we used in those days to avert such tragedies was not in place at Pelly Lake. Our boss, Weldy Phipps, had seen FIDO runway lighting in wartime Britain and improvised such a system at some northern bases where we operated. He saved several lives with this system in Resolute on Cornwallis Island in later years. It is a means of heating and lighting a runway approach by burning fuel. The sudden heat raises the dew point and dissipates the fog briefly. The flash marks the runway. The British system fired the fuel through runway border piping. Phipps simply stored a few barrels of gasoline along each side of the approach end of northern strips and one or two across the end. When the pilot was heard circling overhead in the fog, we would rush out and punch holes in the barrels with an axe and ignite the gas. The pilot would start his approach above the fog layer on the runway heading. As he approached the smoke, poke his nose down into the clag and hope to see the fiery inferno long enough to get lined up and land.

Although Weldy could come to the rescue of other pilots in fog-related distress he sometimes became a victim of the same kind of weather He was ferrying a replacement Mosquito to the operation at Pelly that summer from Ottawa and his ETA Churchill direct from Ottawa was just about twilight. Knowing Churchill weather, this was poor planning on Weldy's part. When it cooled off that evening the fog moved in off the ice pack in the bay. About 100 miles out, Churchill radioed that the weather was zero-zero at his destination. He had no approach aids in the aircraft and even if they had been installed at the airport, ILS and even a radio range or NDB approaches were of no use to him. The Pas Airport was available farther south, but he didn't have enough fuel to get there. Thompson had not been built. Gillam came later as part of the mid-Canada line or hydro power development. Winisk was too far. The only landing site for a wheeled aircraft was a 2,800-foot, very narrow grass strip at Ilford. And it was surrounded by trees and had poor approaches over the trees and hydro wires. It was on the railway line and when the word was passed down the line that this little village was to be the landing place for a Mosquito at night, everyone turned out to watch — all 50 of them. It had been used normally only by the Mounties in a STOL Beaver to take out clients. Cars with their headlights on lined the approach end and the two sides. No-one wanted to park his truck at the far end, as it takes 4,000 feet to land a Mosquito in the daytime and he only would have 2,800 feet at night!

The Phipps magic worked again. With only minutes of fuel left he landed successfully — a feat that called for a lot of luck plus some divine assistance. As if the Phipps miracles would never cease, the next day a 50-mph wind came up right down the runway and, armed with one barrel of fuel, he took off and flew the Mosquito to Churchill. Any other Mosquito pilot trying either to get into or out of Ilford would have been a statistic. That same aircraft, CF-HML, later became an acquisition of the Calgary Aviation Museum.

On another occasion years later, and completely unrelated, Weldy and I were in a rented cab with a macho Latino driver, going from Villavecencio in the Ilanos area of Colombia west over the Andes at night to our operational base in Bogota. We had been visiting Rocky LaRoche, who was based in Villavecencio and who later married a lovely girl from that town and brought her back to

Weldy Phipps, seen here with his wife Fran, was greatly admired by fellow flyers. Fran, who shared Weldy's spirit, became the first woman to visit the North Pole.

Canada. As our cabbie climbed into the mountains through elevations of 12,000 feet, the driver had to keep getting out, opening the hood and adjusting the needle valve in the carburettor to compensate for the thin air. Although the driver did not understand English, Weldy insisted on explaining how he could rig up a mixture control that could be manipulated from the driver's seat just like an aircraft. His technical mind never stopped working.

As we continued, the road clung to the side of cliffs on one side with the other side dropping off hundred or thousands of feet. Crosses along the road marked the casualties that had resulted from drivers going over the cliffs. Our driver, in an attempt to overtake someone ahead of us, got lost in the cloud of dust and veered off the road. We were stopped only by a boulder under the car and left suspended with the front end of the car over a canyon. As the dust settled, the cab driver scrambled through the back seat, climbed over us and escaped out through the back door. His front doors offered only a first step of a thousand feet! Phipps and I, immersed in a bath of *agua diente*, (the local firewater) were close behind him. I am sure to this day that if Phipps had not been in that cab, the driver and I would not be here today. Weldy had a magic about him that served him well.

Churchill is different today. The whale plant is closed, but the bay is still full of whales. You can fly over the deep blue of Hudson Bay, when the ice is out, and see small white specks of Beluga everywhere. The Polar Bears still terrorize the natives every winter. They are shot with tranquilizers, caged and flown out of town, only to come back again for another ride. Wheat shipments through the port have declined and face oblivion due to the very short shipping season. The military and commercial value of the airport has also declined with the advent of long-range jet transports. Tourism is being encouraged, but one ride over that railway, which floats on the muskeg, is enough for anyone. You don't need to go back the next year, and I haven't really told you about the blood-guzzling mosquitoes!

ALAN MACNUTT

One Burning and One Turning

...and my crew walked home

The label Rolls Royce is synonymous with class — aristocratic automobiles, driven by uniformed chauffeurs, filthy rich passengers sequestered in splendid isolation from the Great Unwashed standing in awe on the curb. Rolls Royce automobiles are used as taxis in Hong Kong, but even those are not in great use by bush pilots. The secret of the car's smooth engine is reputed to have been first drafted by Henry Royce on a cigarette package, a procedure not unfamiliar among bush pilots. During wartime Rolls Royce made thousands of engines to power a wide range of aircraft, both fighters and bombers, and in addition to luxury cars made permanent inroads into aviation. Many modern jetliners are Rolls Royce powered and the name still means class, but in the fan jet it now has much competition from quality-conscious engine manufacturers. Spartan made use of Rolls Royce engines after they tested a series of aircraft on photo survey operations.

To get away from the cumbersome Venturas with their altitude limitations, Spartan bought several P38 Lightnings. Those aircraft could carry out high-level photo operations above 30,000 feet and provide wider coverage, at greater speed, with smaller crews, and eyeball navigation. The Venturas necessitated a Shoran triangulation method of photo control. The weakness in the P38 was that the aircraft often went unserviceable, an awkward situation in the far north where maintenance facilities existed only in the engineer's toolbox. They were a rather fragile, effeminate type of airplane that needed to be babied and operated from improved airstrips. By improved, in those days, I mean better than rough gravel, muskeg and dry riverbeds. The P38 was a beautiful airplane to fly, but not a moneymaker. The Allison engines were trouble prone.

The next step up for the company from an economic viewpoint, and a step down for the pilots in discomfort and austerity, was the purchase of a fleet of de Havilland Mosquitoes. These were high-speed, high-performance, long-range airplanes that could take a beating and still perform. British designed and Canadian built, they were made of wood not metal. But the more reliable Rolls Royce Merlin V-12 engines on the Mosquito had a life expectancy of only 400 hours of operation. One experience with Rolls Royce engines that nearly shortened my career occurred in the Arctic in the late 1950s.

In 1954, Weldy Phipps, Russ Hall and John Deacon, Spartan executives, took a float plane and went searching for a suitable site to build a Mossie-friendly airstrip needed for the following summer's photo contract. They found an esker or sand bar near the Back River, not far from Pelly Lake. With ingenuity and imagination they felt it could become a base for our Mosquito mapping operations. The following spring, at break-up, I flew Husky CF-BQC into Pelly Lake with a cook, food, tools, tents and mosquito repellent - no relation to the airplane. Next day a Canso arrived with its belly full of parts, which included a small Oliver tractor with a blade, something like what we would call a Bobcat today. The Oliver worked day and night for about one week and levelled the worst hills off the top of the esker and, in a poor man's sort of grade and fill operation, levelled the soft wet sand for a length of about 2,000 feet and width of about 50 feet.

The next step was to try the strip to see if it would support a wheel aircraft. As a great admirer of Weldy Phipps, I was the first to volunteer to accompany him on the initial flight into Pelly Lake International. We got airborne off the adjacent lake in the Canso and lined up on what looked to me like an elongated postage stamp and a dangerously small piece of sand on which to land a huge airplane. The initial impact was more like a crash and I remember being dashed down into my seat. I didn't see much until we were stopped. Weldy was taller and probably better motivated and could probably see out. It is not true that I was hiding, I just was afraid to look out.

The Canso is a very rugged airplane and can take a severe beating on land or water. She survived the heavy impact of that landing in a sand pile. Even Phipps admitted that we should extend the strip a bit before we tried a take-off, as we couldn't turn to get off the other way. The following week, after more work with the Oliver, our first visitor arrived overhead. Max Ward's new Bristol freighter came in with a D6 Caterpillar tractor and landed safely. We 'walked' the Cat out of the Bristol and extended the strip. That night the much lighter Bristol went back to his base in Yellowknife. We were in business, with the making of an airstrip - two landings, zero prangs.

Within a month, DC3s were bringing in fuel and spares and we were operating Mosquitoes off the sand strip successfully. When you consider how long it takes today to plan an airstrip, do engineering, environmental, financial, and wildlife impact studies before the sod is turned, the forty-five day birth of Pelly Lake takes on more meaning. Granted, it was not the world's best base, but we had 5,000 feet of reasonably level wet sand, about 100 feet wide. It had gas barrels for runway markers, a cookhouse, HF radio for long distance and VHF for line-of-sight communication, fuel, oil, bunks, lots of film for the survey cameras, a small photo darkroom for test strips, and thousands of square miles to photograph and millions of mosquitoes to swat.

Cold was not the big problem in Pelly Lake, in fact the cooler the weather the more firmly packed the sand became and the easier to manoeuvre on the ground. Sand ingestion was our biggest problem. After landing, we had to set the take-off trim at exactly the position we would need to fly away after the next take-off. On take-off and taxiing the tail cone would fill with fine sand seizing the trims for about the first 10-15 minutes of each flight until the sand got sucked out in the vacuum caused by high speed over the tail wheel aperture. Trying to hold a badly-trimmed aircraft level in the climb for 10-15 minutes is a exciting challenge.

ALTIMETER RISING

In those days I recall we had developed a method of getting an Anson started on a cold morning. When we shut down from the previous trip we would preset the trims, the throttles, pitch and mixture all to the start position before shutting down for the night. When it finally started in the morning, after pre heating the engines, it would run at least thirty minutes on the ground with the engines' heat and vibration doing their job before the throttle cables would thaw so that we could move the throttles. If you shut it down at night with throttles closed, or mixture idle cut-off (ICO), she was there until spring or until you found a warm hangar. I recall one occasion flying into Winnipeg in winter with an Anson when one throttle cable froze up. I completed the approach to landing by pushing the mixture control up, then ICO to maintain a reasonable approach speed. The engine went from cruise power to full drag and it was not one of my smoothest approaches. It doesn't do the engine much good either. We learned tricks on one aircraft that came in handy on another, but had never figured on frost and sand causing the same problem only to be solved by the same means.

Take-offs from the Pelly Lake strip were always exciting. We had tried to firm up a parking area by soaking it with used oil. To take-off we would roll out of the parking area at medium power, avoiding a sand swirl that would damage the prop tips. We would then attempt to get our power up as quickly as possible as the speed increased. Often a wheel would drop into a hole and change our direction, or we'd put too much power on too quickly and torque would turn us off the strip and we'd have to dodge the gas barrels with a small slalom. In high performance aircraft like the Mosquito, you can never put the two throttles up together as that would cause a severe yaw off the runway, due to the uneven airflow over the tail surface at low speed. The P38 was an exception to this rule as it had counter-rotating propellers, giving hands-off take-off control.

Take-off power was, as I recall, plus-12 pounds or, in a tight enough spot, whatever you needed or could get, but we could not get the second throttle up fully until we had considerable speed. As the vertical tail surface was small and not far back, the rudder was not very effective because the wheels were biting into the soft sand and the brakes on the airplane were a last resort at best of times.

The British use sea level pressure (29.92 HG) as datum, which became zero pounds on your manifold pressure gauge. Idling on the ground was a negative quantity, i.e., minus 6 or 8. For each pound over static (or 29.92 HG) you interpret 2 inches of boost in North American terms, i.e., plus-12 on take-off was about 30 inches, the static sea level pressure plus twelve doubled, or about 54 inches. We used 3000 RPM on take-off, and it was very noisy. The aircraft would act like a charging rhino, gas barrels would dart in front of you and you would leave crews, aircraft and the cookhouse behind in a sandstorm. If all went well you were soon sitting at 32,000 feet ready to go to work.

My crew navigator was Vince Kluke, a bachelor with a dry sense of humour. He sat beside me. Our camera operator, Barry Cox, sat in a hole near the rear of the fuselage. He was completely enclosed and could only see out through a camera drift sight in the floor beneath his feet. His only exit was through a hinged door that opened outward on the bottom of the aircraft. In other words, if the undercarriage malfunctioned on take-off or landing he was trapped in his little cocoon. No wonder we all wore parachutes. Imagine his feeling on take-off as we careened down the runway, the extreme noise level, sand drifting in everywhere, lurching from side to side, and seeing only a blur of sand streaking by beneath his camera glass. In some of my landings, Barry will tell you it was even worse. On some operations we taped a piece of heavy paper or cardboard over the camera glass for the take-off to avoid damage. The cover was attached to an embedded cord that the camera operator could pull from inside to release the paper after take-off. Single-engine aircraft, although much cheaper to operate, were found unsuitable for mapping as the camera was mounted behind the engine and exhaust gases and engine oil would cover the lens and distort the pictures.

To get back to my Rolls Royce story, the crew and I were doing our second photo trip out of Pelly Lake one afternoon operating at about 30,000 feet. We had just turned for a new line somewhere south of Cambridge Bay. To start a photo line you must get up to stable operating speed and exact altitude, after you turn off the last line. Variations in speed, altitude or heading are not acceptable. In this case, the coast of the continental mainland being our start area, we had to have the camera running 10-15 miles back to ensure the camera operator had time to adjust his drift and interval time between pictures for the new heading and ground speed. This gave the pilot a chance also to check fuel selectors, pressures, oxygen supply, temperatures, and to stretch. To get back on line you were required to hold height to a tolerance of 200 feet and a heading of less than two degrees for prolonged periods without an autopilot or co-pilot. Sometimes you were tired and hungry; you were always cold, or the bends were hurting. A light on the panel signalled when the camera was about to fire. We had to make our minor heading changes between frames or the camera would be tilted and the line ruined.

On the turn I noted a variation in my port oil pressure. It was small but unsettling and so I decided to watch it on the southbound line. When oil pressure drops, you immediately hope you have a defective instrument — NO WAY. The odds of an oil pressure gauge lying to you is very much less than the chance of winning the lottery, and how many pilots do you know who have won the lottery?

As the pressure fluctuation increased, I advised the crew that we were aborting the trip. I was going to feather the prop and head for home. Feathering means the prop blades edges are facing into the oncoming air to give minimum drag. The prop started a normal feather procedure, and then, just as it was about to stop in the fully feathered position, it suddenly ran away. The RPM increased to what was later estimated at 5,000 or 6,000 rpm. The drag from the full fine position of a huge windmilling prop driving a high compression engine is alarming — like pushing a barn door through the air. This onset of high drag happened so suddenly that I was caught completely by surprise. The port side of the airplane seemed to stop and the aircraft pivoted about 90 degrees to the left. The turning movement from one prop in fine pitch dragging and the other at power changed the heading so quickly it almost seemed as if the fin and rudder were gone. The deceleration was so great that the huge aluminum spinner on the left prop pulled its bolts out of the metal and spun off from the prop, feet from my head, and moved ahead of the aircraft. It then came back as the aircraft caught up to it and was shredded by the screaming prop which sent pieces of spinner flying like confetti.

I was terrified. Vince was wide-eyed. But imagine Barry in his cocoon at the back with all the noises and sound effects and no way of knowing what was happening. What had happened was that oil had not been added after the morning's flight and we had simply run out of oil on that side. When the quantity was low I had the indication on the pressure gauge and shut down the engine, but that was not the end of the problem. Normally, a return to base on one engine was no problem at all. What we did not know was that Rolls Royce had not installed a standpipe on their wartime Merlin engines.

As we made our way home to Pelly Lake, with the nerve-shattering scream of the prop in our ears, it was decided to bail out the camera operator as soon as we descended to about 15,000 feet. Our status was deteriorating; parts were being shed from the engine and cowls. I was afraid the engine would seize and the prop would twist off and come aboard and join us in the cockpit. To slow down our speed too much would cause the aircraft to roll. To let Barry bail out at too high a speed could hurt him. We slowed down as much as we dared and Barry bailed out from the rear. Vince looked the situation over carefully and decided to walk home, too. He went out through the floor hatch and earned his caterpillar badge. That left me in sole charge of the shop and feeling very alone. The screaming of the huge prop as its tips exceeded the sonic barrier was terrifying. The engine was burning and, as the situation deteriorated, I decided I'd walk home too.

To control an aircraft in this configuration required substantial control inputs to keep the letters and logo right side up. When I decided to leave, I checked my parachute harness, loosened my seat belt, abandoned the controls and dived for the hole in the floor. Suddenly, the hole rolled up to the top and I was in an inverted aircraft unable to get out and having little desire to stay in. To sort things out I got hold of the control column and worked my way back into the seat to reconsider my options. I had lost my helmet and oxygen mask. Maps were gone. Ten years of debris was all over the cockpit and the aircraft was tumbling, inverted and otherwise, out of control. I had lost 5,000 feet of altitude. It was not what might be termed a smooth operation. I recall tapping out a Morse SOS on the old wartime transmitter key, but that seemed pointless as the crew at base had no way of knowing where the aircraft was. Nor did I. I knew base was about 200 miles away, but I did not even know what direction. The airplane had made various unprogrammed excursions when I was trying to get out. The tundra all looked the same, and I had no maps or navigation facilities. The sun was still shining and that gave me a direction to steer.

Another abortive attempt to hold the wheel with one hand and one rudder pedal with the other, while I worked my way to the lower right side of the tiny cabin floor was no more successful. I lost more height and decided to try and fly it down rather than risk spinning in while attempting to bail out. There is another hatch on top of the Mosquito but it is small, with an axe handle antenna behind it and the fin and rudder ready to dissect any pilot who deserts ship, so that didn't seem like a good plan.

When you fly all the time with a skilled navigator who gives you headings and ground speeds on request, a pilot becomes complacent about his position. In addition, Vince had left me earlier without any discussion of our position. He was in a hurry and I was, as we say, 'unsure of my position'. My ground speed was low, but I found the Back River and followed it back to Pelly with the left prop still screaming in my ears and the left wing was on fire. I was NOT enjoying my day. The base crew heard the noise of the windmilling prop from miles back, and they came out to the airstrip in a Jeep. On short final approach, I made a futile attempt to get the gear down which was probably a mistake. The hydraulics and the left tire were long gone. Most of the wooden structure was burning.

My keenest recollection after a belly landing is getting out the top hatch and running to hide behind an empty gas barrel marker at the side of the runway. I knew the fuel tanks would blow momentarily. The crew in the jeep veered towards the airplane, a flaming mass by now, assuming that my crew was still on board and I had deserted them. When they got to me my mouth was so dry and I was so scared I could barely talk, to assure them the crewmen were somewhere out there in the muskeg, and not in the burning aircraft.

I emerged from all this with the smugness of a survivor and assume the Grim Reaper was not recruiting stokers that day. The best part of me getting back with the remains of the wooden bomber was that I could direct the search crews in a general direction to look for my crew out there on the vast tundra. Only I had the least idea of where to look.

As a result, we found them both before dark and they were home for supper, safe and sound but badly insect bitten. It was a bad day for Mosquitoes. Their rescuer flying a Norseman was Gunnar Inglebitzen,

Al MacNutt (centre) waits at a safe distance for his burning Mosquito to blow up. Left: The fire out, not much remains of HMQ, but a group in Ontario is trying to restore it for display.

later mayor of Churchill. Both earned the caterpillar badges from the Irwin Parachute Co. but de Havilland didn't even ask me for a testimonial!

The problem with a runaway prop on a high-powered aircraft is that the drag on one side is extremely high. You must maintain extremely high power on the good engine to maintain safe airspeed. If you permit the aircraft to slow down you do not have enough air over the ailerons and rudder to control the aircraft and it starts to roll in the opposite direction to which the good propeller is turning. This lack of control might be compared to a helicopter losing its tail rotor. The Chopper pilot, however, has even fewer options. In my situation, it was essential to find a flat, friendly place very quickly and land at high speed in order to keep the numbers right side up.

In today's aircraft there is a reservoir of oil used exclusively for the feathering process. It cannot be used for other cooling and lubrication purposes. As a result of this incident, Carl Burke, President of Maritime Central, had his fleet of Merlin-equipped Yorks modified to install a silo or feathering reserve tank. As there were a lot of York aircraft operated in Canada at that time on the DEW Line project, all the companies using then eventually modified all their engines. Gordon Rayner, a professional engineer and veteran airman, designed and got approval for the modification.

In addition to learning a long-term fix on the engines was necessary, the other fallout from this drama was relatively small. The company lost a $5,000 airplane, a $15,000 camera, and probably $50,000 worth of pictures we had taken. My crew didn't learn much as we were all soon airborne together again in another Mosquito. After my very severe case of dehydration, which is not unusual under the circumstances, my only problem was long term hearing impairment from the extreme noise within a few feet of my head for two hours. Even that has its compensations, as I don't hear now when my wife tells me to go cut the grass!

Pilots, like politicians, are born egotists and will usually try and recall some victory from the chaos. As we said in those days, when airline pilots were said to be limited in conversation to sex, salary and seniority, we survey pilots were restricted to ego, oxygen and high octane, not necessarily in that order.

Aircraft companies today design equipment to fly thousands of hours trouble-free with only minor inspections in a very competitive market. Wartime airplanes and engines were designed for a different role and did not have all the bells and whistles we can expect today. Not even the Rolls Royces.

ALAN MACNUTT

The Brain Game

A vital consideration in air safety

It has long been my theory that many aircraft accidents start in the pilot's mind. The 'snowball' starts with a small mechanical snag, an overload, or badly positioned cargo. A Centre of Gravity problem can result. The weather might deteriorate. None of these problems might have bothered him by itself if he had been mentally attuned to flying the airplane and not preoccupied with brain overload. A bush pilot who can do wonders over a prolonged period operating off northern lakes or rivers can be in trouble very quickly if he suddenly gets a charter into large city terminal during the evening rush hours. The high density traffic has his mind in a spin — changing frequencies, finding the Airport Information Service, sorting out runway charts, getting scolded by Air Traffic Control for being on the wrong frequency. This blizzard of procedures would have been absent in his bush territory. The uncertainty of the busy local area and procedures can set the guy up for a disaster. He could 'blow' a gas tank because he was too busy to change. He could pick up carb ice and not notice it, or he could line up on the wrong runway and cause a problem. A marital problem at home and/or an obscene call from his bank manager can render a pilot unsafe for flight.

I recall doing Pilot Proficiency Checks, during my incarnation as an air carrier inspector in Transport Canada, on pilots who had the required knowledge but who were incapable of doing simple routine exercises in the air because of my presence. Some of these pilots had been operating specific aircraft safely for years. The idea that their career and future was under scrutiny shattered their equilibrium and blanked their thinking process, at least temporarily. During this period of preoccupation, for whatever cause, it takes only a fairly minor occurrence to snowball into a dangerous situation.

On one occasion I was doing a 'ride' at a northern base, on a Fairchild F-27 pilot whom I did not know but who held a responsible position. It was a morning check ride in smooth air. I was in the jump seat and another company pilot was in the right seat. This captain could not fly the airplane. He did not seem to know the basic checks, could not do routine exercises, and got lost trying to find his way home on the VOR. He was given an opportunity to repeat every exercise twice, but to no avail. When we landed, I explained to his chief pilot that I could not pass him. This would cause major hardship to the airline as he was needed on the 'sked' next day. The president of the airline looked me up after lunch and asked me to take the guy up for another ride in the afternoon. I agreed, but I wanted the chief pilot in the right seat.

Captain X showed up and inspected the aircraft properly, did a thorough pre-flight briefing, flew each exercise requested with confidence and precision, did two nice instrument approaches and, of course, passed his ride. The company was relieved and I was confused — why did this man behave differently? Did he have a twin brother? What had I missed? I did not get the answer until several months later when I returned to the base. It seems Captain X was an alcoholic and he had dried out for his 'ride'. He was absolutely incompetent. When he failed - assuming he would be fired - he went and drank half a bottle of vodka for lunch and, back to his normal self, did a good job in the air with no noticeable side effects in the afternoon. The chief pilot had figured this out and fired the pilot soon after the ride. My point was that the guy was operating in an unusual mode - sober - and he couldn't handle the pressure.

The moral of this is NOT to drink and fly well, but the opposite. Operating in a different or artificial environment is a potential hazard as it produces a preoccupation or thought overload that detracts from the pilot's natural ability. Although this preoccupation can happen quickly it is often a long time peaking. Boredom, wearisome routine or complacency can cause it. It is often noted by his peer group but rarely considered a safety factor.

A visit to a shrink once a year, in connection with a regular medical, would, in my opinion preclude many accidents. Like fatigue, we are not always aware of our mental condition and a yearly professional opinion could avoid many accidents. But doctors themselves, who live close to life/death dramas daily, are a high risk as pilots. Like many other professionals who fly mainly for recreation, they have trouble leaving their vocations out of the cockpit.

In my mind, another major problem is lack of proper communication between two pilots in flight. The captain may be preoccupied and miss something, and the first officer may be afraid to speak up in case he is wrong and the captain calls him a fool. Many flight decks are like poor marriages - the co-pilot, like the wife, has the same amount of brains, but the MASTER suffers a Christ complex and cannot be seen to accept help or advice.

The only work experience I had with an airline (apart from 3,000 hours in the jump seat doing checks on airline crews for DOT), was a situation where the airline was joining a union. The co-pilots were all union men and the captains were non-union. As captain, every time I made the slightest slip or bounced a landing, the co-pilot would get his little book out - make a note - and put it safely away in his inside pocket. Not a healthy situation. A constant mistrust of your working partner is very unhealthy in the cockpit.

Scandinavian Airlines, which has one of the best safety records in the business, do a biorhythm pattern on their flight crews prior to assignment together. Two pilots whose emotional or intellectual responses are at low ebb cannot fly together for that pairing. A few years ago, a 747 captain ignored his first officer's protest that the runway might not be clear for a departure in the Canary Islands, and he hit another aircraft killing nearly 600 people. Was he dreaming about something else, or did he just not want gratuitous advice from a junior pilot? 'Junior' in this case is a relative word, as an inexperienced pilot does not get to sit in the right seat of a 747.

A pilot can also get set into a pattern which he later realizes is wrong, but can't think of a way out of the dilemma without losing face. Continuation will compound the problem. Self-confidence and ability to make decisions is mandatory for a successful pilot in command. However, too much machismo can be disastrous and it behoves the most experienced pilots to be aware that they are not infallible and that they will lose less face to admit their mistake earlier and not pursue a bad plan to a dead end. Captain John Gallagher recently wrote, "All pilots make mistakes, but those who live the longest are those who swallow their pride and correct them immediately." The line between self-assurance and blind ego is sometimes very thin and pilots themselves are poor judges of which is which.

The amazing reliability of the new jet equipment makes mechanical problems a minor factor in aircraft accidents. For example, I can recall 25 or more engine failures in flight which required single (or no) engine landings. My son, to my knowledge, has only shut down one jet engine in his career. So to analyze accidents we should now look in the cockpit and study psychology, crew behaviour and co-operation. Does the captain listen to his first officer's opinions? If not – why not? Ego, the fear of losing face, emotion, competition and sky rage have to be examined as well as medical fitness.

The long periods of isolation we survey pilots spent in small desert towns and in the Arctic took their toll, especially on the younger crew members. I only recall one pilot who had to be removed from the Arctic in restraints. His career did not suffer, however. The Department of Transport hired him in another capacity.

One factor that bothers older pilots who have been flying a long time is deafness. I have always said that I can judge an older pilot's experience in flying hours by how deaf he is. In wartime we started out with leather helmets and a rubber tube called a 'gosport' through which the instructor yelled instructions to the student. This improved with bigger aircraft, but not much and not quickly. After the war major airlines issued headsets to crews that cost $2.00 each and that did little to either improve communications or blank out the noise from the high powered engines with the short stacks.

The North Star, a Douglas designed transport that was modified by Canadair with in-line engines, was a noisy brute. Since the headsets did little to cut the roar, crews did not use them all the time. Earplugs were not invented yet. No one told us in general aviation that the DC3 would deafen us if we only used the headset to talk to the tower on take off and landing. Canso was even noisier with the Pratt and Whitney 1830 engines roaring and the big propellers whirling just a few feet away from our heads. The smaller de Havilland 'Otter' was one of the noisiest with a fire wall resonance that was deafening. The purpose was to push the exhaust blast into an augmentor that gave additional thrust, but it was an unkind act for the pilots who flew them. Max Ward, in his autobiography, has a better explanation of this noise factor of the Otter and he should be an expert. Like many others from that era of poor headsets and often up to 10 hours a day of continuous engine noise, I have poor hearing.

A friend, Tom Wilson, flew Otters for years and when we talk we wind up shouting at each other to be heard. This diminished hearing causes us to ignore a lot of chatter on our headsets - a dangerous practice - because inevitably some of that chatter could be directed at us and could be important. This can annoy your crew and company and cause a great deal of frustration. Younger pilots, who have always enjoyed the modern noise dampening headsets, believe the line between deaf and stupid pilots is very thin, and tend to treat us accordingly.

Propwash

Life's little ironies

Many incidents come to mind as I look back over the years. Some are so obscure and lost in time that I wonder if they were fact or fiction or whether it happened to someone else! I cannot blame anyone else for the day I test flew the Ercoupe owned by Pearl Peacock and Dorothy Book. Harry Smith ran an aircraft repair business and one day, after he had done a Certificate of Airworthiness on an Ercoupe owned by the two members of the Ottawa Flying Club, he asked me to do the test flight to complete the paperwork. I lined up on Runway 22 at the north end of Ottawa's Uplands Airport and, as I started to roll, I cranked in some cross wind correction to compensate for drift. Suddenly, I was dodging runway lights and wound up in the grass. I went back for another run and exactly the same thing happened. I went back and explained my problems to Harry and asked him if he had broken the controls on the inspection. I thought they might have been crossed, although they looked normal. Silly question! Harry patiently explained that since the Ercoupe had no rudder pedals, you drove it like a car with the wheel and told me to go back and try it again.

As I had been flying high performance airplanes and the Ercoupe weighed about 1,000 pounds with probably only 75 horsepower, I nearly pranged it on a test flight. Remember, any airplane will bite you if you don't treat it right and read the book first. You can kill yourself just as easily in a Piper Cub as in a military jet, if you don't observe the basics.

--oOo--

Ottawa Flying Club reminds me of the night the local branch of the 99's (an international organization of women pilots) put on a dinner there for members and guests. The story came out somewhat later that when the ladies were preparing the dinner, which was an enormous pot of spaghetti, they had left the plastic tongs in the pasta while they enjoyed happy hour. The plastic tongs melted and when dinner was announced they could only find the top of the tongs - but served it anyway!

—oOo—

On a more chilling note, I recall clearly one night I was flying alone from Inuvik on the MacKenzie Delta of the Northwest Territories to Mould Bay, our Arctic survey base on Prince Patrick Island. Earlier that day, I had completed my 7½-hour survey trip

Al couldn't let his Aero Commander 'cold soak' in the Arctic. He had to press on – almost fatally..

and was just returning to base at Mould Bay when the RCMP called and asked if we would do an emergency evacuation of an Eskimo child to the hospital in Inuvik. Glen Mockford, our engineer, fuelled up the Aero Commander and I was on my way.

Our company wasn't in the charter business, but there was no other suitable aircraft available. The Aero Commander was fast and warm and so we could take the child's mother along. We shut down the survey sortie to make the medical evacuation. In the Arctic in early spring it is daylight 24 hours a day, so our three crews flew around the clock.

ALTIMETER RISING

When I dropped the patient off at Inuvik three hours later, the doctor asked me to wait a couple of hours as I could probably take the sick child back home again. An hour later, just as I'd got into a real bed with clean sheets in the hotel and was getting some much needed sleep, the hospital called. The child and mother must stay and I was free to go home.

I was exhausted, the bed was warm and clean, and this was the first time I'd been between clean sheets for weeks, but duty called. Another consideration, of course, was that if I let the aircraft cold soak at the airport for the eight hours sleep I needed, it wouldn't start again without several hours preheating, and there was no easy way that would happen. A quick call to my family - phones are rare in the Arctic - a cab to the aircraft, the Commander started, and I was northbound.

One of the most basic concepts of flying in the Arctic is the fact that magnetic compasses are unreliable. Even as far south as the MacKenzie Delta, you must use about 30 degrees of variation between the magnetic compass and a true heading. As I knew the route well and had been over the area many times over the years, I did not bother getting out my astro compass. Nor did I take time to steer into the sun and get a true heading. I simply lined up on the runway, made the correction for variation 'on the roll', and headed north. The airplane should have been on survey, and the sooner we got the job done the sooner we could go home. The radio station took my flight plan in the air and I sat back in boredom to face the 650-mile trip, only about half awake. I didn't bother to use my long-range communication set; it wouldn't be needed.

A few minutes later, fortunately before I had flown out of VHF range, a radar controller, who obviously had nothing better to do, called and seemed to want to start a conversation. This is normal in the remote areas as they get very bored, but I was tired and cut him off. A few minutes later, he called again and asked my heading. Once again I cut him off short, as he could establish that himself. He obviously just wanted to talk. The third time he called and started talking, the hair on the back of my neck went up in that old familiar feeling, as he told me I was not heading for Mould Bay; my heading was taking me out over the Arctic Ocean towards Russia. As the landscape was absolutely featureless, just a solid mass of sea ice after I passed Tuktoyaktuk, there could be no map reading on that route. I planned to parallel the shore of

Banks Island about two hours after take off, but we did not have beacons that were useful more than 50 miles or so from the station. Sacks Harbour on Banks Island would have been my first landfall and beacon on this route. I would then fly up the west coast of Banks to Mould Bay.

It became obvious that in my hurry, and by ignoring the basics, I had nearly flown out over the ice pack and would have probably never been heard from again, if that faceless radar operator had not taken an interest in my heading. If he had ignored me when I cut him off and gone for coffee, I would have been out of VHF voice communication in a few more minutes - and I didn't have enough fuel to make it to Russia!

When I took off from Inuvik, I had built in my variation the wrong way and was flying a 60° error in heading. My problem here was I thought all this was too easy and I was not paying proper attention to basics.

Complacency is a killer in aviation.

—oOo—

Norm Spraggs, who was our engineer in Nigeria on most of our winter operations, had completed a maintenance check on a Piper Apache one day and asked Doug McDonell to do the test flight. Doug was a macho type who later found a home in the Department of Transport. As he returned from the routine test flight, Doug saw a couple alone on the beach outside Lagos and decided to investigate. When he dived down and flew along the shoreline, he saw the pair was trying to increase the African population and unwisely came back for a second pass. Their siesta ruined, the couple had got up and were making their way to their car, and when Doug went by the second time he noticed they were white, or as we say in Africa - Europeans. The result of Doug's test flight was a lot of political static for both him and our company, as he had inadvertently disturbed a senior official of the British Embassy and his friend. Doug said it proves the theory that only mad dogs and one other group go out in the midday African sun!

—oOo—

ALTIMETER RISING

Troubleshooting in the air is a no-no. Several years ago a new Lockheed 1011 passenger jet flew into the Everglades in Florida because all the crew members were trying to solve an undercarriage light problem and no one was attending the shop. There are no hills in Florida to fly into, so when the altitude hold slipped out, or was inadvertently shut off, the big jet just eased down and squatted into the swamp. These crew members were not engineers, but were distracted and all playing with a problem and no one was doing what they were well paid to do - fly the airplane. Fortunately, most of the passengers survived.

I still believe it is an advantage to know as much about the equipment you are flying as possible and you can be a better, safer pilot if you know how the machine works and why. You can also be a better engineer if you can test fly the machine and see how it works in flight. But above all, do your flying in the air and your troubleshooting on the ground.

An airline pilot's action when a component fails in flight is to reset the circuit breaker once. He then should go to an alternate system and have his co-pilot write up the snag for the ground crew to fix that night. Our American friends call such a malfunction a 'squawk.'

ALAN MACNUTT

The Kenya Caper

Trying to take high tech to Africa

In 1957, Spartan Air Services purchased a failing business called McLaughlin & Co. in Nairobi, Kenya. The objective was to secure some of the large aerial survey contracts known to be coming up in Africa. I was sent out as operations manager to try and put it all together. My wife and family came later and we spent two years in that beautiful country with its magnificent climate and serious political problems.

McLaughlin was a charming, if ineffective gentleman of the old school and a veteran of the Rhodesian Air Force. He had several aircraft at a base at Wilson Airport, a suburb of Nairobi on the edge of the game park. The new airport, Embakasi, came much later. The McLaughlin inventory consisted of an Airspeed Oxford, which was a twin trainer used by the RAF at one time, a Percival Proctor, a Gemini, and some antique cameras with which he had collected many debts.

With a friend, the family takes a mini safari on a Kenyan plain.

ALTIMETER RISING

Knowing little of the sleeping giant that was Africa at the time, Spartan was naive about the inertia established by Africans after years in the sun. We introduced a Mosquito, equipped with a shiny new, state-of-the-art Wild camera from Switzerland for high level survey. We brought in pilot Ralph Burton from Winnipeg and Vince Kluke, photo/navigator from Ottawa, to crew it. Dennis King was its maintenance engineer. Next came a Bell helicopter, a first for the area, and Jim Lapinsky from Calgary to fly it. Then we acquired an Apache and the Piper dealership from the Placo distributor in South Africa. Finally, we imported two spray aircraft to work on the vast cotton fields in the Sudan and on farm crops in the highlands near Mount Kenya and in the Rift Valley.

Other members of this hodgepodge survey company included Terry James, born in India and later emigrated to Canada; a photographer from Australia; two lab technicians from Scotland; a Sikh engineer; a Sikh carpenter; a Portuguese accountant from Goa; and two smart young women in the office. Muriel Jennings married the local salesman for crop control chemicals. The other was married later to the head of the flying police wing 'Punch' Bearcroft who had achieved fame in fighting Mau Mau from the air in the earlier attacks in Kenya. He was an excellent pilot, which was unusual because he had lost an arm and flew with one hand! I remember trying to persuade my superiors in Ottawa to let me hire him to fly the chopper when Jim Lapinsky went home - maybe that is when they knew for sure they had sent the wrong man to Africa - a one-armed helicopter pilot yet! Africa is different, but Spartan brass in Ottawa told me you need three hands to fly a helicopter, two is bare minimum.

One of my worst mistakes was made in the selection of spray aircraft. The Stearman I bought from Russ Bradley of Carp, Ontario, was a good idea. It behaved beautifully in the high elevations and pilots all vied for the pleasure of flying it on spray and dusting operations. Many of the farms were at the 8,000 and 9,000 foot level and the Stearman with its Wasp Jr. 450 hp did a great job. One small problem gave the pilots a thrill on take off. There was no baffle in the spray tank and as you started your take off run the liquid would all rush to the back of the huge tank. Then, as speed increased, it would surge ahead again and develop a cycle. The pilot, usually working from a short, rough strip, was busy trying to compensate for the bunny-hopping take-off run and deciding whether he could lift off on this wave or if he could afford to wait for the next one. Airborne, it was fun to fly.

Acquisition of a Taylorcraft Topper seemed like a good idea at the time. But Al lived to regret the choice.

The other spray aircraft we bought, which I admit was my idea, was a Taylorcraft Topper. We already had lots of junk in Kenya but I knew Taylorcraft always built good aircraft and this one was sitting at the factory brand new. It was one of only seven built, I believe, and we could have it for about $7,000 U.S. How could I refuse? But it was a dud. As it was my decision to buy it for Spartan (Kenya) Ltd., I felt that I had a responsibility to make it work. On one operation we booked a spray job near Mount Kenya, planning to use the Stearman. But as the Stearman was away on another job I set out with a load of dust in the Topper to work at a farm over 8,000 feet above sea level. The farm had no airstrip, which was unusual as most large farms had a strip for the flying police during the days of the Mau Mau battles. With a full load, it took me a long time to get up the hill. When I finally found the farm, the men with flags who were to act as markers had got bored waiting and had all gone for 'elevenses'. When they got back and we got the job done, I was low on fuel. I had been at full throttle since take-off and knew I would not have enough to return to Nairobi. But in a stubborn effort to make the Taylorcraft pay its way I had forced on, regardless. I got out of the highlands and started back to a strip I had used previously to refuel. About five miles short my destination the engine sputtered out.

The scourge of Africe, the desert locust can make a mess of an airplane.

I arrived in a cornfield, sans dignity or further plans. It was the only time I had ever run out of fuel. I paid off the white farmer, lined up about 20 Africans with machetes and, in about two hours, I had fuel and a take-off path. The aircraft was not damaged but the president of our Kenya company was curious when I returned to Nairobi as to why the trip had taken so long. He was an English lawyer, specializing in divorce cases and was, fortunately, not too conversant with aircraft.

Another offshoot of this operation that I enjoyed was a spray contract for the Desert Locust Group, an international organization based in London. Desert contractors provide aircraft and crews to await the influx of the dreaded locust, which arrive annually from across the Red Sea and the Gulf of Aden. The insects, which measure up to four inches in length, eat their weight several times a day. They would devastate grass, cotton and corn. Like the goats, they even thrived on the thorn bushes prevalent in the area. They ravaged all the crops as they swarmed through Djibouti, Somalia, eastern Ethiopia down through Kenya and even into the Rhodesias. We got a contract for the Topper to be based at Hargeshia in British Somaliland and flew daily sorties to track the vast swarms.

I would cruise to places with names like Jigiga and Dire Dawa, up to Asmara in Eritrea, south to Mogadishu and Kismayu in Italian Somaliland, and to small towns in the Ogaden. The geography was fascinating, wild and lonely, with drifting sand and only the nomadic tribesman on camel trains. There was an occasional oasis with a collection of traders stopping briefly to trade, fight and refresh their camels.

The main export at that time was hashish, the popular but generally illegal smoking drug. Camel drivers would appear on the airport ramps with loads of hashish or 'choat', as it is know locally, to be weighed and loaded on the Ethiopian Airlines Convairs for transport to hard currency countries. The camel trains moved at night to get their load of hash to the airports for the early departures. There was some rule that these narcotics could not be transported openly in the country although they were grown everywhere. But it was legal to fly the dope out of the country.

This story would not be complete without trying to explain the shock a pilot feels when a locust storm is encountered. On patrols you may be suspicious of what seems like a storm cloud ahead -- unlikely because relative humidity is about ten per cent and rain is rare. You enter the 'cloud' usually at a hundred feet or so as patrols are at low level. The noise is incredible. It's as if you'd been bombarded with millions of ball bearings. The windshield goes opaque instantly, the leading edge of the wing gets a layer of honey-like goop and all the screens specially designed for the air intakes, clog up in the front. The cure is to pull back immediately on the pole and at about 100 to 200 feet you fly out of the swarm. The terrifying noise stops but you still can't see out. Next you get on the horn and alert the base so they can send out airborne sprayers to destroy them as they leapfrog over the countryside.

The Topper soon paid for itself in long patrols. One day, on one of these patrols, I decided to follow the railway. I was having fun sitting just a few feet above the crew of a freight train as they hauled their cargo over the desert. The train crew was sharing the fun but suddenly seemed to get very excited and pointed ahead. I looked up just in time to see the train was about to enter a tunnel and I got out of there pronto.

My most vivid memory of the locust operation was getting blood poisoning while working on my spray tank one night. We knew little of toxic chemicals in those days and when I cut my arm I was not too concerned. I took seriously ill one evening and I woke up two days later in the grubbiest hospital I have ever seen. My treatment was pills for malaria. As most white Africans immunize themselves by a constant intake of gin and quinine water, they are well protected. However, it was assumed someone with a crude North American accent would not have been exposed to enough quinine. In any case, I survived, but the African locust nearly got his revenge.

The Kenya Company folded in a few years. Once again, modern technology was proven ineffective in overcoming African inertia. The resources are there, both in people and in the ground, but the secret of getting it all together is still obscure.

The long way home from Africa gave the family a great education. Shaded from the Sahara sun, Cairo was one of the lessons.

One of the many benefits of the two years in Kenya with my family was the trip home. We found the school system bizarre, the teachers knowledgeable but brutal, and the children intimidated. One school told us it took three teachers to hold down our five-year-old son to give him his needles. It nearly needed three more to restrain me from punching him out. The British system of sending your kids away to a boarding school until they grow up and get some maturity is a difficult concept for me to understand. It seems like they consider their kids a nuisance, that someone else will get paid to look after them, leaving the parents free for parties and golf and an uninterrupted social life.

Returning to Canada, after it was decided we could not correct all the problems in Africa in two years, we decided to take the tourist route. All my previous trips at this time had been by airlines, usually a Pan American Super Constellation out of New York via Dakar, Senegal, Roberts Field, Liberia, and Lagos, Nigeria, into Nairobi, or via London and through Rome and Benghazi in Libya and Khartoum in the Sudan and down the Nile in a Viscount.

We decided to supplement our trip home and sail from Kenya to Europe and fly home from there. We boarded the Union Castle in Mombassa and cruised north through the Indian Ocean, around the Horn of Africa and through the Red Sea to the Suez. An overland trip took us to visit Cairo and the pyramids, which our children greatly enjoyed. We rejoined the ship at Port Said. The Mediterranean was not calm and sunny as advertised, but neither were there enemy jets overhead. Our children enjoyed seeing the islands of Stromboli, Elba, and the monuments to Christopher Columbus when we docked in Genoa, Italy. They enjoyed learning the history of all the spectacular sites.

From Genoa we visited Rome, drove through the Alps to enjoy the beauty of Switzerland. At this time in our lives we had not yet seen the majesty of our own Canadian mountains and thought Switzerland must be the most beautiful country in the world, and probably the most expensive. Both assumptions were wrong.

After some terrifying experiences driving in Italy and through the Alps, we were still not prepared for French drivers. We turned the car in and finished the tour on the famous Orient Express. Our kids were unprepared for the mixed sleeping arrangements at bedtime on the train, when men and women of several nationalities all got undressed communally and got into their beds. Nor was my wife ready for the many-handed Italians in Rome, where the men would make a fuss over the kids and get a quick feel of her at the same time. Or the day in Paris when she was shopping for perfume and came back out to find a lady of the night propositioning me in front of the children in a shopping area in daylight.

We often forget, as Canadians, particularly the westerners, that we are a new country. Quebec can boast of a 400-year history, Vancouver only one hundred, but the history of Europe goes back to Day One. It is hard for North Americans to be humble because when we see the French plumbing, the English food, the German arrogance, Italian political fiascos, we have reason to believe that we colonials must be doing something right.

On my return from Kenya, Spartan was winding down. Contracts were harder to get as many small companies were getting into the field, most of them operated by former Spartan employees. My services were no longer required and for me the glory days of Spartan were over. Our company had too many high paid executives resulting in a high overhead that was not competitive. Our main customer in the geophysical division, Canadian Aero Service, was flirting with Bradley Air Services in nearby Carp and soon Bradley was beckoning to experienced crews. Many of us changed flags of loyalty and flew for a lower paying company, with minimum overhead, rather than get an honest job where we would have to work nine to five.

The joys of survey flying were not yet over. Rather like a driver faced with a red light, it is easier to turn right and keep going into the unknown instead of stopping and considering where we were going or what was our objective.

At several points in my career there were airline opportunities for experienced pilots, as the cyclic shortage of pilots happens about every five years. From an economic viewpoint, I would have been well advised, in retrospect, to have taken an airline job in the late 40s or early 50s, but I was impatient to get into the real action where I was captain. Airline starting salaries were traditionally low and routes routine and predictable. Senior airline captains retiring today have attractive pensions, but I note many of them have grey hair and medical problems. Finally, I would have had to quit commercial flying eight whole years earlier had I been with an airline, and would have missed the joys of flying for the questionable pleasures of the golf course.

ALAN MACNUTT

The Bradley Experience
A handshake was a done deal

At the end of the war, Russ Bradley came home from air force service as a pilot and started a flying school in his hometown. Unlike many other such schools, his survived for about 20 years. Its successor continues today on the same wartime field at Carp, Ontario. Bradley Air Services went on to form the genesis of First Air, operating scheduled services into the Arctic with Boeing 727s from Ottawa. But that progression became possible only after Weldy Phipps, became a partner with Russ. Weldy had been the ingenious pilot/engineer who had risen through the ranks of Spartan Air Services to become chief pilot, maintenance superintendent and chief operations officer. He had nurtured a desire to go deeper into the High Arctic and develop more practical air transport there than had been possible with Spartan. When Spartan withered, Weldy moved to Carp and went to work on his vision of the north. Bradley's established a charter service based at Eureka on Ellesmere Island and later from Resolute on Cornwallis Island.

The partnership with Bradley lasted only a couple of years and broke up after a brief and stormy difference of opinion over Bradley's personal business ethics. Phipps left Carp and formed Atlas Aviation, which he based at Resolute Bay on Cornwallis Island. Through hard work, determination and exceptional skills in that inhospitable climate of blizzards in winter, mosquitoes in summer and government interference year round, he made his million. He was also smart enough to leave the Arctic and spend it in more exotic climes like Prince Edward Island, Bermuda, Texas and throughout Caribbean while he and Fran still had the good health to enjoy life on their 48-foot yawl. Like Max

Ward and a very small group of natural leaders, Phipps made aircraft do things that everyone knew was impossible. They opened new areas for lesser men to try and follow.

One summer in the 60s, I took a contract with Bradley to provide an airborne limo service to three geologists who were exploring the rocks on Banks and Victoria Islands. Another pilot, Bob O'Connor and I were given a PA12A each to fly the scientists around and land them where they needed to do their studies. It was an operation designed by Weldy Phipps while with Bradley. It was a service made possible by his modification of the light aircraft for practical exploration. Our small single engine Piper, used mostly for agricultural spraying, had a double seat in the back for two passengers or cargo. That space had accommodated the bug spray but now was to carry the rocks the geologists chose to study.

Weldy Phipps (left) holds a flight planning session on the tundra with scientist of the Geological Survey of Canada. His aircraft modifications cut costs and greatly increased their survey territory.

To understand scientists, it helps to assume they are a little crazy to start with. They examine rocks that were formed millions of years ago, but have difficulty with the more immediate concerns such as food, shelter and survival. Hundreds, of them huddle in, cluttered offices in Ottawa for nine months of the year. There they are hounded by government bureaucrats and budget cuts in order to spend three summer months studying fossilized seashells, caribou droppings or the aurora borealis. During the ensuing nine months, of course, they write learned papers that only another PhD can understand.

Geologist Dr. Ray Thorsteinson checks a muskox graveyard on Banks Island, NWT. His Phipps Super Cub was flown by Al MacNutt

In my one and only summer serving as an airborne cab driver to three of these over-educated people, I had only one shower. This was when I had wheedled, whined and begged them to let me go south to Cambridge Bay after two months to get mail, supplies and to phone my family. I recall one chap would regularly, after a 12-15 hour day out on a transit, come in, take off his boots and put socks, unwashed for a month, on the leg of ham for a footstool. This same guy, when we caught some fish for a diet that did not include ham, would fillet them, throw the fillets away and make a stew of the fish heads. Ray liked to be one tough cookie in the field. He has spent many summers in the Arctic and is recognized as one of the north's top explorers. Dr. Ray Thornsteinsen had more degrees, papers and books to his credit than James Michener. John Fyles, another of the geologists, could see the humour in Ray's eccentricities. He also had a brilliant career in science, but did little to help improve our living standard during our summer on the Arctic islands. Wanting a bath, a normal meal, mail, or family contact all seemed to confuse these experts and jeopardize their studies.

ALTIMETER RISING

Each morning I would board our Super Cub and, with spare gasoline on board, set off on a transit with one of the scientists. We would make ten to fifteen landings a day on eskers, muskeg, riverbeds, gravel piles, anywhere that it appeared safe. The high performance Pipers were equipped with the huge, low-pressure tires developed just for this purpose by Weldy Phipps. While the geologist collected samples of rock and flora with his pick, I would read a book, make a cup of tea and mainly figure out how in hell I could take off from there. Often I would fly a test take-off. If it came out OK, I'd land back at the same place. If not, I'd find a better area nearby and wait for my passenger there. In an hour or so, the geologist would return carrying rock samples all tagged and identified, ready to fly on a mile or two and land again at another outcrop. These sample bags were small cotton sacks each loaded with a few pounds of rock. By night, when we returned to our primitive camp, the airplane would be well over gross. Doing this continually all summer, hitting boulders and hummocks all day long on take-offs and landings was hard on the airplane and cracks soon developed in the undercarriage cluster. Cracks also started appearing in the pilot. I know the scientists were as unimpressed with my performance as I was with theirs.

Our sleeping bags were filthy from filthy bodies, our tents were a disaster, and the mosquitoes defied explanation. Head nets and gloves helped keep the insects off to some degree. The front of our tents was black with swarms of them. Unless you could face into a wind the insects were intolerable. Even body functions like urinating were difficult and dangerous, and trying to empty your bowels while avoiding a mass attack of mosquitoes in critical areas was a challenge. One system that was uncomfortable, but worked best, was to walk into a lake and squat, lower your pants and try to evacuate your bowels under the surface in the water, which was only a few degrees above freezing.

That was not the best summer in my 50 years of flying, but one I'll never forget. Helicopters have replaced the Super Cubs as transport in the islands and tons of rock samples are still brought home to Ottawa each fall for winter study.

On one of our transits around the north end of Banks Island, we discovered a large muskox graveyard. Hundreds of carcasses, mainly skulls, were lying on the ground. Whether this was a mass slaughter or whether the Eskimos carried them there, we never established. Scientists have been studying this find for many years and are still attempting to find the reason for this mass kill.

Muskoxen are interesting. When we flew around a herd they would form a circle to fend off their imagined attacker, hooking the calves back into the circle with their curved horns. An ugly, shaggy beast, they wander the wilds and seem to migrate among the islands, over the winter ice pack looking for better grazing. Perhaps they were using the same guidance system as we were. Navigation among these islands is easy when you have mastered the concept that the sun is south at noon and appears to move 15 degrees clockwise each hour. In cruise, the simplest way was to turn the aircraft into the sun every twenty minutes or so, set the suction-operated gyro and turn back on heading. For example, at 9 a.m. the sun is on a heading of 180° minus 45° or 135° true. At 3 p.m. it is at 180° plus 45° or 225° true. As your map is in degrees true and not magnetic, you can apply your gyro setting directly. An Astro compass is more accurate if you have current tables and a reasonably stable platform. It can be used as well at night. A sextant is handy to establish your position when on the ground, but cannot be used by a one-man crew while airborne.

You could forget beacons or VORs in those days, and especially forget the magnetic compass. When near the magnetic North Pole, which in those days was near Bathurst Island and is now much farther north, the magnetic compass is unreliable. Another concept that fools many people looking at the Arctic wastes on a map is the size of the area. If you stop to consider that the meridians of longitude all converge at the true North Pole, then you will appreciate that these islands are much smaller than they appear on a flat map projection. When flying from Frobisher Bay (Iqaluit) to Alaska, for instance, you are going through time zones and meridians of longitude at a great rate.

Long months in the Arctic in semi-isolation with no mail or phone service is hard for a family man and harder for the wives at home raising the kids. Every time we heard of an aircraft accident in the North, we would all try and get a message out to our wives that it wasn't us, we were okay and would be home soon.

ALTIMETER RISING

Russ Bradley was a shrewd businessman, but he had a great loyalty to his pilots and expected that loyalty returned. After completing my Arctic contract with him, we had a breast to breast in the back of his hangar and I was promised better jobs in future. A handshake to Bradley was *done deal* and, unlike many aviation bosses, he always remembered his promises. Instead of paying his pilots a regular salary, Bradley preferred to keep survey crews on a contract basis. When we were away working we were well paid, had good parts support and usually a good production bonus. The advantage of this for Russ was that, apart from paperwork, when we came home we were paid off and his overheads decreased. Then, when he got the next contract, he would rehire only the hungriest and the most productive crews. He had no obligations to anyone.

And so it came to pass that I was soon on my way to South America operating a Beech 18 on magnetometer survey out of Maquitea Airport in La Guaira, Venezuela. La Guaira is the port for the city of Caracas, a city of over three million people consisting of about 10 per cent very rich people, 70 per cent very poor people, and the rest who scramble in between. Driving from La Guaira to Caracas through a small gap of the Andes, a distance of about 20 kilometres, you could see some of the homes of the poor. They literally live in holes in the ground, or *favelas*, which are caves in the hillsides. Generations of Venezuelans live in these caves with no water, heat or sanitary facilities. They forage for food wherever they can find it. Small wonder the crime rate is very high and life is cheap. Sadly, massive landslides early in 2000 killed thousands of the favela dwellers.

Venezuela is the California of South America, the melting pot, the adopted home of machismo, oilmen, and adventurers. Many are from Europe and other parts of the world who found their home country uncomfortable because of shady business practices, politics, bigamy, or whatever. The indigenous people are farmers, fishermen, and miners, of a mixture of races - Caribs, Arawaks, and Spanish. They are emotional, hot headed and mostly Catholic. The population has grown from six million in 1958 to over 20 million.

The rich people live in suburbs like Altamira or Florida (meaning flowers). These residents are usually prominent in business and politics. They have lots of inexpensive servants, luxuries, automobiles and magnificent homes. The cost of living is very high and the few who can afford it enjoy a very fine lifestyle. In many major cities our employers put us in a good hotel, making it impossible to make a fair judgment of the people and the diverse culture of a country. Only when we worked in the interior of the country, based at the airport in a small town, could we see the real people up close.

Our contract was to do a geological survey for a large mining company to establish where the lush overburden of rain forest hid the mineral wealth of the interior. There are large bodies of ore mined in Venezuela to feed the smelters in the U.S.A., Europe and Japan, and the search is always on to replace the reserves. Oil is also abundant in Venezuela in the Maracaibo Lake area in the north. The country is a major player in OPEC, the Organization of Petroleum Exporting Countries.

As you can imagine, Canadians working in Venezuela required a strong local helping hand. Wisely, we chose Jacques as our expeditor when we set up our survey camp in Cuidad Bolivar, along the Orinoco River in the State of Bolivar, 650 kilometres south of Caracas. Gringos trying to run any kind of operation in such environments met many problems unless they had a good 'agent'. Jacques was a well-educated local, originally from France, who was equally at home using French, English, Spanish, or the local dialect. He was dark and slim with a good sense of humour. He knew everyone in the area, their weaknesses and how much we had to pay the authorities for food, lodging, fuel, landing fees, transport, and protection, so we would not wind up as somebody's hamburger. Such skill lent credence to the rumour that he was evading the French police. He was, however, very comfortable with his local wife and family. He found an old truck we could rent and drove it for us although there were few roads. Licences and insurance were not a consideration. Jacques did a great job to accelerate our work.

On the operation with me were fellow veterans of aerial survey Dwayne Olson and Hal Mordy. Ossie Darrick of New York City was the fourth crewmember. Our engineer was Russ Bradley's brother, Clayton, and we flew the Bradley Air Services flag on this operation. Also on site at La Guaira was a Kenting Aviation crew from Oshawa, Ontario, working on the same project operating a Lockheed Lodestar.

ALTIMETER RISING

Our aircraft was a war surplus T11, one of the many configurations of a Beechcraft 18, a twin-engine trainer used by the military as a trainer. Behind it we towed a magnetometer over the rain forest on predetermined lines or transits, looking for spikes on our recorder which would indicate mineral deposits. Another instrument gauged the overburden on top of the ore while a third device measured the area for radioactive minerals such as uranium. This information was all compiled at night in our makeshift labs and synchronized with our strip camera, which was always running when on survey. It does no good to find the mother lode of raw ore in the rainforest if you can't prove where you found it. The developed film was used to confirm the accuracy of the track.

Our routine was for one crew to fly from daylight at 6 a.m. to noon. The other, after a brief stop for refueling, flew from noon to dark at 6 p.m. It is important not to return to an unlit airstrip after 6 p.m. in the tropics because when the sun goes down it goes STRAIGHT down. Unlike our northern latitudes, the equatorial region is pitch dark a few minutes after sundown. With no alternative airports nearby you have huge and immediate problems. (Aircraft have been known to return with spears stuck in them and the poison on those spear tips is for real!)

Pilots are by nature very competitive. Each crew would try to get more miles of productive survey done each day than the other crews, much to the company's delight. The sooner the job is completed, the more the profit. Often the opportunities of daylight were stretched to the limit.

The type of flying involved was very boring over the jungle and savannah, but as maps were very primitive we had to work quite hard to stay on the transit lines and this kept us busy. Most survey aircraft had extra fuel tanks installed for endurance, but in the tropics it was always important to quit survey lines on time to return and land before dark. The locals had better things to do than sit around after 6 p.m. with the runway lights on and radios open waiting for some Gringo pilot who is not back on time. You had better be self-sufficient. You would get little support from non-flyers to whom the concept of an aircraft lost or out of fuel over the jungle at night was not a major concern. Even our mechanic, the owner's brother, left the airport one night at five minutes after six, when we were late getting home. He wanted to go for supper.

Clayton didn't really understand the concept and when we got back a few minutes late, we had no transportation and couldn't even get out of the locked airport. When we eventually found our way back to the hotel, he got an animated briefing on his responsibilities to his crew members.

In such instances, pilots tend to try to understand the gulf in philosophy between themselves and non-fliers and realize we live in a different world - not necessarily better, just different. A flyer, believing one of his buddies might be missing, overdue, or in trouble, would make all possible efforts to establish contact. That could include bribing airport officials to stay open, declaring an emergency, or whatever it takes to help recover a colleague. Ground crews in Latino and African countries, regularly shut down and headed to the nearest bar or restaurant.

I recall coming into Las Palmas in the Canary Islands, off the northwest African coast, one night at dusk. About 30 miles back, the radio at the airport went off the air, the beacon shut down, and in the distance we saw the airport lights go off. We were on a flight plan and would have arrived on schedule, but the airport workers wanted to go home! We were on our own, and the nearest airport that would be open was in Casablanca, Morocco, or Dakar, Senegal, either of which was about 700 miles away over the Atlantic Ocean. It was a clear moonlit night and, fortunately, we had no trouble finding the runway and landing. But it teaches a pilot to be wary, as the weather, even in the beautiful Canary Islands, is not always clear.

During the Venezuela operation, where the south end of our lines took us to the Brazilian border, we broke line one day and went to see Angel Falls, the highest waterfall in the world. The water drops over half a mile, twenty times the height of Niagara Falls. Billy Angel's airplane still sat near the top of the falls which bears his name. He was an aviator/explorer who mapped that area in the 1930s. The aircraft has since been moved to an aviation museum in Caracas

Venezuela is cattle country and many of the residents, like Albertans, relate to cattle. In Caracas our crew used to dine at a nice little steak house called Mi vac y yo, which translates to 'my cow and me'. One of the owners' regular acts was the 'shoo through'. A couple of steers were chased through the restaurant, across the dance hall and out in the general direction of the kitchen. This happened during the course of every evening and was enjoyed by the patrons, especially from a safe distance. One evening a couple of military policemen were dining there unaware of this time-honoured procedure. One of the steers paused at the officers' table and munched on their salads to the delight of the crowd. As if in payment, one dropped a resounding splat of manure on the floor which splattered the officers' uniforms. As the military were not particularly trusted or admired in the country at that time, the crowd burst into boisterous cheering and applause. One of the officers, feeling his dignity had been violated, pulled out his revolver and emptied it into the steer. After a lot of thrashing in agony the hapless beast died on the dance floor. The officers left, the steer was dragged out, the floor mopped, and dinner and dancing continued. Such was the lifestyle in Caracas.

The Latino thrives on noise. As driving habits are somewhat like the kamikaze drivers of Italy and France, the authorities made a ruling that no one within the city limits could use a car horn. That was rough on the drivers, particularly the cabbies. As their windows are always

Crews in Caracas relax between flights over the Venezuelan jungles.

open, they developed the habit of pounding the outside of their car doors as they screamed at other drivers in the never-ending traffic jams. Every cab had a bashed in left front door!

Another incident, which tells a lot about the high rollers in Caracas, occurred at the end of our job in Venezuela. We had all worked hard and the job was done early, below estimated cost. Our customer, Venasca decided to have a party to celebrate and thank the crews. Venasca was a consortium of Americans and Venezuelans using cheap Canadian labour for the dangerous part of the job, not unusual in big business or in wartime. They convened us at the biggest hotel in Caracas, one of the International chain called the Tamanaco. They plied us with the best of food and spirits on a sunny Sunday at the poolside. As the group ran to 20 or 30 including the compilation and office staff of the company and their groupies, the bill must have been in the thousands of dollars. The American boss, called for the *quinta* (bill) as dusk approached, tipped the maitre d' and signed the bill. He used a fictitious name and room number, as he did not live in the hotel. He told us all to leave quickly. Such is business in Venezuela.

Survair Ltd. Comes of Age

Apaches to big airplanes

In 1964, Canada planned to conduct a vast geophysical exploration program in Western Canada. To the existing survey companies that smelled of money. But Spartan was no longer a major contender and Bradley did not have a geophysical data compilation department to put the information together into geological maps. That left Canadian Aero Services in line for some very lucrative contracts. However, in a coup worthy of a new African republic, five of their senior executives met privately and formed a clandestine company called Survair Ltd. and made a low bid on the proposed government contracts.

The new company sprang onto the scene with several years of magnetometer survey work in Saskatchewan, Alberta and the Northwest Territories. When the lawsuits were settled, Survair survived and then had to hire pilots and crews and buy aircraft. Not wanting to make Spartan's mistake of high overhead, they bought two new Piper PA23 Apaches and a used Cessna 182. It was an economical purchase of aircraft that had a minimal capacity for the job. I had just returned home from a photo contract in Chile flying a P38 for Aero Service Corp. and was immediately propositioned for a job by the new company. Working out of Ottawa, it would appear I might spend more time at home. But Survair did not want to pay a fair salary. Their equipment and organization spelled cheap, and besides, they did not want the expense of engineers or aircraft maintenance. Their policy was to only hire pilot/engineers. As the company grew, they could not sustain this policy, as there was only a handful of people with both licences. On a photo survey aircraft, a pilot/engineer (one man doing both jobs on his aircraft) might fly only a few hours daily and have all the cloudy days to do maintenance. On a geophysical survey, pilots often fly eight or ten hours daily, seven days a week, and have little or no energy left for proper maintenance of the aircraft.

It seems these five executives, some of them Americans, had learned the power broker business well from Uncle Sam and, as I had a big mortgage ($65.00 per month), I was persuaded to be the first pilot to join Survair Ltd. Ed Kozystko, an old friend from many previous jobs, was next. It was no secret that Ed and I were the first two hired because we both had engineer's licences and maybe were hungrier. Soon others such as Ken Atkins and Ken Fraser joined after they returned from overseas jobs. Based on solid bread and butter government work, low overhead, and their choice of the best crews, Survair soon prospered. They bought Aero Commanders to replace the Piper Apaches, and later DC3s and Cansos for overseas work. The company had been started with very little capital from salaries and small personal loans, and so these purchases were all from profits.

The first summer there were only two pilots, Ed Kozystko and myself, and we worked hard. We both flew a lot of hours and I recall when we came home in the fall my new aircraft looked very second hand with oil leaks and wear and tear. Ed's looked like a factory demonstrator as he was much more interested in maintenance than I and he was a much better engineer. This may or may not be why I was named chief pilot, which allowed me to hire some smart young people. Tom Appleton and Peter Mitchell, had just gotten their commercial licenses from the Ottawa Flying Club and proved good choices. Tom eventually went on to become a vice president of de Havilland in Toronto and later president of Canadair's amphibious division. Peter and I worked on many jobs. We flew together in Conair and side by side in Airspray, bombing forest fires for many years. He also had a career in airlines flying 737s out of Yellowknife, as well as long contract periods overseas, and a stint on Arctic resupply in four-engine transports. Other pilots to join Survair were Jean Gaudry, Gordie Carter and Gus Morneau, all from the RCAF.

Former Spartan pilots who joined us were Rocky La Roche, Stu Hill, Ken Atkins, Syl Panagapko (remustered from electronic technician), and Doug McDonnel, who later joined Department of Transport and retired early to go hunting and fishing. Dave Croll, Ed Jensen, Bob Gant, Andy Lambros, Bob Pilgrim, Ken Dempster, Bob Veale, Pat Korpatt, Terry James, Jerry Deluce and Bob Beaton also joined at various dates. Among our engineers was Glen Mockford, later a Boeing 727 specialist in Edmonton; Norm Spraggs, later a de Havilland service rep in Toronto, retired from Bombardier, in Brisbane, Australia; Dave Dorosh, who spent many years in Colombia, later in Edmonton; Dave Rowlands, who joined a helicopter company in Newfoundland; Bill Doherty, Esko Patovirta and Red Sarsfield retired. Ed Kozystko now lives in Whitehorse, Yukon. Rocky La Roche, Gordie Carter, Dave Croll, Bob Veale, Les Hanna and Jim Neals all died in crashes of survey aircraft.

Survair, at its best, lacked the glamour and pizzazz that made Spartan famous. Spartan was a hangover from wartime. Survair was out to make money. Spartan was a pilot's company, developed, owned and operated by pilots. If you were a good pilot you had it made. Survair was owned by shrewd businessmen, and geophysi-

Tom Rowlands (right) accepts Survair's first big aircraft, a Canso, from John Bogie of Laurentian Air Services Limited, in Ottawa.

cists. Pilots to them were a necessary evil. I remember shocking the directors one day at a meeting where we were planning to move an Aero Commander that was finishing a job in the Arctic Islands to its next job in Scotland. They wanted to bring the aircraft home to Ottawa first and then ferry it to Scotland. When I pointed out that the airplane, where it was in the Arctic, was closer to Scotland than to Ottawa, and that we would double our ferry costs to bring it home, they wouldn't believe me. I went and got a globe and proved my point quickly. Tom Rowlands was smart with figures, but lacked the knowledge on how to deal with pilots. He did not know how to use our ego to his own advantage. Confrontations were a way of life. I taught Tom to fly on our Cessna 182.

Murray Turner, a smart young engineer who had come to the company from a railway-engineering project in Labrador, was sent to Africa to set up a base in Lagos, Nigeria. Murray did a great job for the company, secured much survey work from government and industry, and provided us with a well-organized base of operations to work from in Nigeria and surrounding areas. When we arrived in Lagos in the fall, with aircraft ferried in from elsewhere in the world, Murray would have all our survey maps ready, the customers and bureaucratic problems under control, transportation and drivers laid on, and we could get right on with the jobs. Previously, so much time was spent on overseas survey jobs doing clerical and expediting jobs that frequently the aircraft sat idle for days while we got a work permit, or a shipment of supplies out of customs, or found suitable accommodation. Needless to say, the crews were delighted to have Turner doing the expediting. He had a knack for logistics and got the best production out of his crews by treating them like people, not like necessary nuisances. He was a 'field man' himself and could relate to our problems and helped solve them quickly.

I recall visiting Senegal once to see Peter Mitchell, Syl Panagapko, Esko Patovirta and crew, who were working a Canso on a magnetometer survey in an interior town called Tambacounda. On arrival, I could not find accommodation in Dakar because the First International Negro Arts Festival was in progress, headed, I believe, by Louis Armstrong. The celebration of art in a huge slum city struck me as unusual. I finally found accommodation on board a Russian hotel ship anchored in the harbour. Food was normally beet borscht. Huge hulking policewomen swarmed over the ship and intimidated the guests. The Vodka was good and plentiful. I only stayed there two nights until I could get a flight inland, but every morning I was glad to see we were still tied up in the harbour. I had a fear the ship might move out to sea at night. We had weird ideas of what the Russian people were like at that time.

When we left Dakar one morning on a local DC3 to fly to Tambacounda, the old Senegalese aircraft lost an engine shortly after take off. As it was very hot and we were fully loaded, the pilot had his hands full to keep it flying. The cockpit door had been removed, so all the passengers added to his troubles by rushing up front to see what was wrong. Returning from Dakar later that gweek, I was on a Russian Tupelov en route to Ghana. Curious, I visited the cockpit and noted an empty Vodka bottle under the captain's seat. The first officer actually went back to the refrigerator in the passageway and got himself a cool beer to drink during the flight. It was a hot day! The inside of the aircraft was filthy and there were only a few seat belts that worked, only Russian was understood on board, and most of the emergency exits were blocked. Russia had a major hold on several African countries at that time and I'm sure, like Cuba, they are suffering withdrawal symptoms today as Russia can no longer afford to develop colonies due to her financial problems at home.

These were good years for me. I enjoyed ferrying aircraft to different countries, watching young crew members develop. When I had ferried an aircraft to its new base of operations, I would wait to see the crew got a good start on the job, then go home for the next assignment. Although I travelled constantly, I was never away from home long. I had a chance to teach my kids to fly and watch them mature.

I learned to produce training and operations manuals for our pilots, and gained some office savvy and knowledge of administrative paperwork that would serve me well in my next job, with the Department of Transport. I also secured Air Transport Board approval and an Operating Certificate for Survair to conduct charter operations, invaluable later when survey waned and Survair became involved with First Air, a charter operation in the Arctic and Antarctic using Twin Otters and later Boeing 727 airliners.

As a person's importance and job description develops in any company, it seems to increase the banana peels dropped by your peer group. We had a single engine Otter on mining survey work in Northern Ontario one winter and the pilot, Jerry Deluce reported that his engine had high oil temperature and low oil pressure - a bad combination. I immediately went out to see him and flew with him. The Otter was on skis and flying a crew over the barrens of Northern Ontario in winter on a defective (single) engine, was not good business, so I recommended returning the aircraft to Ottawa to find the problem.

From the safety of his warm office, Bill Doherty assured us there was nothing wrong with the engine. The president hastened to agree. He had just paid the bill for the overhaul on the big Pratt and Whitney and was positive the problem would go away. Besides, we pilots were all a bunch of prima donnas. And so: ultimatum time. In a pilot-oriented company my position was correct, but Survair was a money making company and, as the president explained, I had been crying safety too long and he was happy to accept my ultimatum and once again an abrupt career change was imminent. In good executive fashion, I had trained my successor, Stu Hill, who had been promoted to chief pilot the year before when I had been moved up to Operations Manager.

I believe I had helped Survair move their flight operations into an efficient, well trained, and disciplined organization. All our pilots were now instrument rated. Stu and I had check pilot authority to do instrument rating renewals on our crews overseas, and we had proper manuals and training on a regular basis. I completed 16 Trans-Atlantic ferry flights with various aircraft, often by myself, and felt I had made a substantial contribution to the company's bottom line. The flying contribution was not recognized as I was often told they could get better and cheaper pilots anywhere and, of course, they did. Cheaper, of course, was easy, but I was proud to have worked with these crews who were tops in their field.

ALTIMETER RISING

One of the pleasures I had during my Survair years was working with some fine people - other than line pilots - who taught me a lot and who I will not forget. Grant Mervyn was an electronics engineer, a confirmed bachelor and computer wizard. Dwayne Olson, my magnetometer operator on many jobs is a life-long friend. Norman Spraggs kept our aircraft safe to fly on prolonged overseas operations. Reg Kensey came to survey from the RCAF, where he was an engineering test pilot and was a great source of help and comfort. He was a very experienced senior pilot, a top-flight professional engineer, and a gentleman to boot. Reg left to head up Sanders Aircraft originally in Montreal, then they acquired a hangar in Gimli, Manitoba, to convert British built de Havilland Herons, a light 4-engine aircraft, to the turbine powered ST27 and 28. Neither Reg nor the Sanders Aircraft survived, as his health declined and the Sanders could not compete with the Twin Otter and other mass-produced turbine-powered twins of the time.

An example of my admiration for Reg stems from an incident that could only be settled between pilots. One day we needed to move a Canso to Africa. The airplane had a huge antennae system on the nose that would be very expensive to remove and replace after arrival in Africa. I said it had to come off. We would not have been able to maintain height if we hit icing flying through the Arctic. And we had another impasse -- pilots versus bean counters. Reg, a senior officer in the company, accepted my invitation to come see for himself and that night, after work, we took the Canso up for 15 minutes with Reg in the captain's seat. He came back convinced I was right and my decision ruled. You can talk to pilots; bean counters are a race apart. Eventually, Survair formed part of First Air and their main customer, Canadian Aero Service, had their ranks weakened when several of their executives split to form Terra Surveys.

Sometimes, after a long or dangerous flight, there is a period of mental fatigue on landing, when the noisy engines are shut down and the objective has been achieved or the crises averted. Such situations are often alive with drama real or perceived. The transition from the 'air world' to the 'ground world' can be very traumatic. The pilot might have been working at fever pitch, possibly bringing back a sick aircraft. Mechanics, relatives, news media, customs, police or petty bureaucracy wants a piece of him. The drama over, the option to sit in the airplane quietly and relax the mind and senses, is not available to the pilot. He is expected to get out smiling and be prepared to answer inane questions from people from another lifestyle who have no concept of the tremendous emotions he has experienced.

This seldom happens to airline pilots. Fortunately, their lifestyle is scheduled to be routine with no surprises and they are continually connected to the 'ground world' via one radio tuned to company frequency for continuously updated information. Another connection is maintained to Air Traffic Control for traffic data and instructions, and another receiver available for ATIS (Automatic Traffic Information Service) at the destination airport.

In flying we frequently get mixed signals. One instrument will point in one direction yet we know from all our experience and instincts that the opposite is true. Hard decisions must be made quickly and acted upon NOW. There is no protracted analysis, no committees or sleepless nights to make the decision. The pilot does it and makes the commitment. We take responsibility for our lives and the lives of others and most of our decisions are correct - or we are not flying any more. Small wonder then, that pilots are individuals, self-confident to the point of arrogance, isolated in their thoughts to the point of being anti-social, and relating mainly to their own peer group until they are sometimes considered snobs.

This aura of self-deception is fortunately punctured periodically every thousand or so hours of flying, by making a very wrong decision at some point and getting into serious trouble and having a good scare. This creates a philosophical balance and produces some maturity - and some grey hair. Younger pilots, elated by their sudden elevation into the ego-inducing status of commercial flying, are at the greatest risk. They have yet to have their share of terrifying moments to challenge their god-like status and inspire a handy alternative plan. My wife's analysis when boarding an aircraft is not scientific, but very astute: "Make sure the captain has some grey hair."

Ferry Flights

The long and lonely route to Africa

Survair Ltd. had several large geophysical and photo survey contracts in Africa in the 1970s and one of my jobs in autumn was to ferry the aircraft from Ottawa to the contract area. I would remain for a few weeks to get the crew started and then come home by airline. In the spring, I would repeat the process to get the aircraft back to Canada for the summer season. This fleet varied, but was usually two aircraft to Nigeria each winter, often a DC3 and an Aero Commander, and others in North or West Africa.

I won't dwell on the beauties of Nigeria, but it has great natural resources. Lagos, the capital, had at that time over three million people living in one of the biggest slums in the world. Whereas most western cities have a large area of respectable homes and a small slum area, Lagos had a vast area of ramshackle hovels and a tiny area of nice homes called Victoria Beach where the diplomats and rich merchants lived. Nigeria itself, the size of British Columbia, has a population of over 112 million people in 1995- four times the number in Canada.

The DC3 and Canso ferry flights were fairly routine, but one winter I had to position an Apache in northern Nigeria. The cameras and magnetometers were removed to make way for a large gas tank in the cabin. The Apache has a normal range of about 700 miles but the longest leg on the trip - Newfoundland to the Azores - was about 1,500 nautical miles. With all this extra fuel in the cabin the aircraft was, of course, very heavy - much above legal weight limits. Such flights required clearance from our insurance company and the Department of Transport. The problem for the pilot was not just getting all that weight airborne, but keeping it airborne in case he lost an engine during the early part of the flight. For the first 600 miles, a power loss on either engine would put you in the water very quickly. After about 600 miles the airplane might maintain flight at a reduced speed if you sat down on the air cushion just above the waves. But the trouble could start very early in the flight. Immediately after lift off, the pilot could get a surprise. The huge load of fuel would cause a severe pitch-up or tuck-under if the centre of gravity had not been calculated correctly. The midfield area at St. John's Airport in Newfoundland has several burned out areas caused by ferry crews who erred in this calculation and had their ferry flights terminated in the first mile.

With various scenarios running through my mind I exercised my brain by thinking of all possible disasters -- icing en route, electrical failure killing my primitive navigational aids. How would I find those tiny islands in the vast Atlantic? My nerves were somewhat frazzled as I arrived in St. John's, Newfoundland, on my first Apache trans-Atlantic ferry flight. Later, trips became routine, but the first Apache ferry trip was, to me, a big deal.

In those days, there were weather ships holding known positions in the Atlantic issuing accurate weather conditions 24 hours a day. All this volume of weather was compiled by people who had professional pride in their work. There was less of a cultural and professional gap between the civil servant forecasters and working pilots. We had the feeling that the weatherman cared. Pilots in return reported every hour, updating the weather they were experiencing. With computers, radar and unions this mutual trust and co-operation has, I fear, decayed. We pay far more now for an inferior product from faceless computer operators who seem to have little pride or interest in their weather forecasts. We pay vast sums for sophisticated radar systems, which are merely toys if the operators are not motivated.

In St. John's, about 3 a.m., I called the met. man who reported acceptable conditions with favourable winds, and so I checked out of my hotel and called a cab. As I would not get to eat for many hours, I decided I should have a good meal. I asked the cabbie to take me to a restaurant. Unlike airline pilots, crews of small airplanes learn to eat and use the facilities whenever they get a chance. Those are options we do not have in the air in light aircraft.

The cabbie took me to a restaurant in the city, but it was closed. No problem. He knew a truck stop on the highway that would be open. With my raw nerves (and irritable personality), I made a few snide comments about why he hadn't taken me there first if he knew it was open. He let that pass. The truck stop was closed as well and I boiled over, giving him a mouthful of garbage about driving me around running up the meter for a big fare and ripping me off in general. I was badly nerved up in anticipation of this flight, and here was a handy target on which to vent my wrath.

The cabbie listened to me for a while then said calmly, "Come Wit' me bye. I knows a place that's open fer sure". A few minutes later he pulled up in front of a very modest home in a residential neighbourhood and told me to come inside. He then went upstairs in what turned out to be his home, got his wife out of bed (about 4 a.m. by now). She came down in a housecoat, all smiles, to cook breakfast for her husband's 'new friend'. The house was immaculate, the hospitality was unparalleled, and the homemade bread, heavy molasses and hot coffee she produced will never be forgotten. I grew up a little bit that morning.

I have met and enjoyed the hospitality of many people in many countries in my years as a survey pilot in Africa, South America, Canada and Europe, but never have I met people like the Newfoundlanders. They have few of life's material things, but have a wealth of spirit and charm that makes them happy with their lot and willing to share their pittance, often with the people who do not deserve such treatment. And, probably best of all, they have the wit and words to express themselves. The accent may be difficult to follow and phrases unique, but their jokes about Mainlanders far surpass any jokes heard about the Newfies. After that lesson in humility, the ferry trip to Africa was a cinch, which sets me up with a chance to expound a theory that I have had for many years.

Professional pilots normally have two medicals per year, both with physicians who look at our body, give us an ECG, check eyes, heart, urine, etc. No one ever looks between our ears where most problems start. I believe many accidents could be averted if one of our annual medicals was by a psychiatrist who dug into our mind and could get us to lighten up on a lot of accumulated pressure that builds up in there. In my opinion that pressure causes us to spend too much time thinking about personal trivia, where we should be left free to concentrate on our job.

Alone in an Apache, low over the Atlantic, Al checks his heading with an astro compass en route to the Azores.

Self portrait

The story of the generous cab driver who bought me breakfast is supported by a survey taken in 1991 that showed Newfoundlanders contribute more per capita to voluntary social services than any other province in Canada, and on a lesser income.

When I started doing regular solo Trans-Atlantic ferry flights, I had some reason to be nervous. Not long before, a senior Spartan Air Services crew had lost a Canso during a ferry trip and later at least two good friends had gone down during Atlantic ferry trips. Rocky Laroche was killed going into Iceland and Sam Taylor went down somewhere in the Atlantic. Years before, Sam had been ferrying a Canso to Europe with Jock Buchanan as his navigator. At some point during the long overnight ride, Jock is alleged to have told Sam to alter course ninety degrees to starboard. Now, a nine-degree correction without a good reason (like receiving a beacon ahead) would be a major error, but a ninety-degree alteration must have produced some panic on board. Just before they ran out of fuel, the story goes, Sam saw lights ahead and thinking it was a ship, decided to land near by in the stormy sea, hoping to be picked up. It was, in fact, a lighthouse on the rocky shores of Ireland. Lost at night with no airport, Sam landed and ran the Canso up on the rocks to get his crew out safely onto terra firma. The airplane was wrecked but the crew was safe. Buchanan, being a thrifty Scot, remembered in the confusion that he had left his wallet on board and went back into the wrecked Canso in the dark to try and find it. While he rummaged in the aircraft the wind or tide changed. The Canso slid off the rocks and out to sea. The hull had been badly damaged and would not float. Jock was never found.

In later years, insurance companies would not insure the company's aircraft until the Captain had made three successful crossings, so being a veteran by then I could train people like Pete Mitchell, Gordie Carter, Tom Appleton and others.

During the 1950s Spartan imported a lot of war surplus aircraft such as Mosquitoes and Yorks from Britain. Peter Knox of London, who ran a professional ferry service, did most of the ferry work to Canada. Why Spartan did not use Don McVicar of World Wide Airways in Montreal I do not know. Don was well set up for this type of operation but perhaps the 'foreign is better' mentality was in place.

The story of Max Conrad, the preacher who ferried new, single-engine Pipers and Cessnas from the factories to Europe and Africa, has been told frequently by reliable scribes, so we assume it is at least basically true. Once when Conrad lost his oil pressure over the Greenland ice cap, he landed on top of the snow pack on wheels at about the 10,000-foot level. He walked over to a nearby cliff and spotted a ship below. Somehow he is alleged to have hailed the ship, which by pure coincidence had just finished its survey mission and was setting sail for Boston. Conrad was back in the U.S.A. in three days, untouched even by frostbite, in time for a scheduled dinner speech! Few ventures in the frozen wastes end like that and the logistics of that story give me great problems. Conrad is a very famous man in aviation history and I do not mean to be cynical, but in my experience, life is not really like that, especially in the Arctic. But of course I've never been granted any divine assistance. The rumour goes that Mr. Conrad carried a Bible instead of a flight manual. He claims 116 Atlantic crossings in light aircraft before the ice cap incident. By the way, there are no cliffs on the ice cap in Greenland. It has a surface like a vast mushroom, beautiful in the sunshine, but treacherous in dull light as you can't gauge your height over the ice pack.

Farley Mowat's recent book The Farfarers tells an interesting tale of time when Iceland and Greenland were fertile, inhabited countries and the climate semi-tropical. Conrad just arrived a little late for such hospitable landing facilities.

In all, I completed 16 trans-Atlantic ferry trips with aircraft ranging from DC3 and Canso down to Piper Apaches and Aero Commanders. The butterflies usually disappear after two or three hours over the ocean, and, unlike the wartime ferry service, there was never anyone shooting at us on the European end.

One of the nicest views a tired ferry pilot can have is the tip of Pico sticking up through the clouds at 8,000 feet as you approach the Azores from the west. It means that you have an emergency landing base (U.S.) on your left and only an hour or so to go to Santa Maria, food and a bed.

One ferry trip I will long remember was from Johannesburg, South Africa, to Accra, Ghana, with a Cessna 182. We were to do an aerial photo job for Kaiser Engineering for the Volta Dam in Northern Ghana. From Johannesburg it would seem logical to fly up the west coast of Africa through Southwest Africa (now Namibia) and Angola up to the Cameroons then west through Nigeria. The west coast of Africa however, is over 1,000 miles of barren country with sand dunes to the sea and no civilization (at that time) or airports, fuel or accommodation.

This area is referred to as the Skeleton Coast in a recent National Geographic article. In fiction it is the Diamond Coast. By any name, it is most inhospitable to transient pilots. We opted to go farther north into Rhodesia (now Zimbabwe) and across Africa east to west at a very undeveloped latitude. En route we ran into the inter-tropical front and had to choose a place to land in an open area of eastern Angola rather than turn back over the jungle. The big hazard here was not the four-legged animals and zero-legged crawlers, but the curious two-legged natives. I don't believe the ones who crowded around our aircraft wanted to eat us, but they wanted everything else in our airplane. I took off much lighter than I had landed. They had probably never seen an airplane before at close range and probably never seen white men. After a few hours on the ground with dark approaching I felt I was safer airborne lost somewhere in a thunderstorm than to spend a night in this company. I took off and headed west for the South Atlantic Coast.

That night in Luanda, the capital of Angola, we stayed in a hotel across the street from the prison. The shrieks and screams of the prisoners, which the concierge told us was 'only the guards beating them' is something I'd like to forget. The legacy of Portuguese colonization is not good reading as Angola and Mozambique will prove.

Africa is a large continent and as each state has probably a dozen tribal languages as well as the European language inherited from colonial times, it is sometimes hard to communicate. A kitchen knowledge of French, English, some Spanish and some Swahili, may sound impressive, but Tanganyika (now Tanzania) is based on German. Angola and Mozambique are based on Portuguese. Zambia needs Flemish. Everywhere one needs 'baksheesh', or whatever the local word is for handouts or bribes -- the universal language. A good supply of American one-dollar bills would do more to expedite a ferry flight than a tail wind.

The Cessna worked well up the coast and, after the intertropical front, I followed the coast through the Congo (Zaire), Gabon and Rio Muni and overnighted at Douala in the Cameroons. It was distressing to learn that a Scottish Caledonian crew had just flown a charter DC6 into Nikongsamsa, the local mountain, killing everyone. That is the only real mountain in that part of Africa and I guess they must have been lost in cloud. We landed in Port Harcourt the next day and met some Canadian pilots who were there working as float pilots on Cessnas for the oil companies in the delta of the Niger River.

From there we went on to Ghana, where we were to spend much time later on contract. Along this Gold Coast my favourite stop was

Abidjan, Cote d'Ivoire (Ivory Coast). Abidjan the capital was then the Paris of Africa. The industry, styles, boutiques and fancy stores outclassed Johannesburg or anywhere else in Africa. The hotels were a treat. As colonists, the French married into the local tribes in Africa and the prosperous local blacks had all the dignity and haughty demeanour of their ancestors from France. The old hangars were still there where the P38s and other aircraft were delivered by ship from U.S.A. and reassembled. In the late 90s, tribal warfare had erupted again. France is not welcome and the beautiful country that produces 40 per cent of the world's cocoa was in chaos. In Ghana, we based in Accra and Takoradi, once a base for wartime ferry crews.

Robin Stanton, a young South African who was to operate the camera in Ghana, accompanied me on this trip. He knew much more about handling Africans than I, or else we might have wound up in someone's pot. He, of course, spoke Afrikaans, which was very helpful on the first part of the trip but was better forgotten in the all black portions of our journey. South African whites are disliked when they move farther North into the newly formed countries where the blacks hold political power. Robin emigrated to Canada and later settled in California in the electronic business.

Ferrying aircraft around the world is an interesting business. You learn about various weather patterns, trade winds, and weather system reversal in the southern hemisphere where all the highs and lows turn the opposite way from what we are familiar with. Different stars are used at night for navigation. Polaris, the navigator's star from the beginning of time, is not visible below the equator. Most airports were closed at night in Africa in those days. Finding a forced landing area at night was not a happy thought. We seldom flew much after dark.

The African coastline, generally speaking, is a wide, smooth, sandy beach that would make an acceptable forced landing area in many areas except South Africa. In Kenya the Flying Police regularly used the beaches all along the coast to land and refuel and to hunt down poachers shipping out rhino horns to the Orient to make the Viagra of the day and elephant tusks to supply Europe with ivory for luxury items – all for profit. At low tide the beach was flat and firm and suitable for light aircraft.

In ferrying we also had to have available the various radio frequencies of the countries we traversed. Add that that we needed the correct currency for services and bribes for customs officers. Many countries insisted we hire a customs agent -- often an off-duty customs officer -- to process our paperwork. Such transactions cost a lot of money. Strangely enough, many African and South American countries would honour an International Shell or Exxon Fuel card. Not so strangely perhaps, I once had to pay cash before they would put fuel in my DC3 in Texas.

Ferrying aircraft never becomes routine. What is firm policy one year will be completely different with a change of government the following year. On one occasion I was going into the new international airport in Lima, Peru with a P38 Lightning. They had just installed an Instrument Landing System localizer, which is a UHF electronic beam down the centre of the runway to guide bad weather landings. Although I did not have an instrument in the airplane to receive the localizer, I was charged $75 U.S. cash to use it. I was told someone had to pay for the installation. Then you could try explaining that on your expense claim. Traversing through Panama with a Mosquito I once had to make out 24 copies of the same form to satisfy the authorities.

In many undeveloped countries the use of English as demanded by the International Council Aviation Organization, is merely a theory. When radio communications became too difficult it was sometimes appropriate to turn the radio off, change frequencies or ignore the caller and hope they did not have aircraft with strange markings to send to chase you. This was never a problem over the North Atlantic as communications were excellent on HF (High Frequency) channels; available by extending our trailing antennae and tuning the transmitter for maximum peak on published channels. We could talk to Oceanic Control out of Newfoundland for about 1,000 miles and then usually could change over and raise the Azores or Scottish Oceanic for weather updates from Europe. On the South Atlantic there were no ships or communications of any kind and in Africa and South America often too much chatter, much of which was not understandable.

In U.K., radio communications were always excellent and professional. You could fly into London's Heathrow or Gatwick Airport with a bare minimum of chatter. Prestwick in Scotland and Shannon in Ireland were also good stops with little hassle, but Paris, Rome and Lisbon were bad news. Unlike airline pilots, we had our choice of where to land and always went where we would have the least hassle and quickest turn-arounds.

For the sheer joy of flying it is hard to beat the thrill of completing an Atlantic crossing in bad weather and hearing the radio beacon coming into range on your receiver straight ahead going into the Azores, or see the peaks of the mountains poking through the clouds as the sun comes up after a long and lonely night over the stormy Atlantic.

The Aero Commander arrives in Lagos, Nigeria. From left: Al MacNutt, Grant Mervyn, and Ed Jensen.

The Longest Night

Brazil to Africa at 100 mph

In October of 1965, Survair had a Canso (CF-JJG) operating on a geophysical contract in Venezuela. The Canso had many advantages for such operations. Initial cost was low; engines, props and spares were plentiful. The aircraft had large tanks designed for a 'loiter' mode over the ocean for anti-submarine operations during the Second World War. This also was an advantage in its role as a survey platform. It was capable of long trips without refuelling, even though its cruise speed was slow.

Initially, the survey Canso towed a 'bird' or magnetometer head on a long steel cable about 200 feet below the aircraft. The sensor was reeled back in before landing and had a guillotine type of shear that could sever the cable quickly in the event of an engine failure or other emergency in flight. This mechanism was awkward and heavy. Later the bird was built into a phallic type 'stinger' mounted in the tail cone. The Canso was also used extensively for electro magnetic (EM) survey looking for exotic minerals. This consisted basically of a large transmitter in one wing tip sending a strong charge to the earth. A receiver in the other wing tip received and recorded the bounced signal. The time delay was recorded and, combined with tracking cameras and recording machines, the geophysicist could pinpoint potential ore bodies and determine how much overburden to expect. From these data the commercial potential for ore, carbon deposits or other interesting bodies of natural resources could be calculated.

In addition to its fuel capacity (over 1,400 imperial gallons), the Canso wing was a rigid structure with minimum flex in rough air, unlike the modern airliners with their exotic alloys. The rigid wingspan of over 100 feet allowed the survey aircraft to position the transmitter and receiver that far apart for EM work. In some applications, a large wire rope antenna was coiled about the outside of the airplane in the form of a loop secured at each wing tip, at the nose and at the tail. From above it looked like a small baseball diamond. This increased the drag and slowed the aircraft slightly, reducing the single engine capability. The worst fear we pilots had, however, was accumulating a load of ice on the loop. As drag increases as the square of the speed -- and we never had much speed -- it did not greatly affect performance. But the weight of ice on the wire loop could break the attachments and the steel cable could come in through the cockpit and through the props. Also, in cutting holes for cameras, navigation aids, and drift sights the belly of the airplane was no longer watertight and would soon sink if we had to land on the water. On one occasion one of the survey Cansos being ferried to South America went down in the Caribbean off Venezuela. I'm told they had to rent a boat and haul the aircraft to a beach in a hurry before it sank.

The Canso was big and ugly, and slow and heavy on the controls, but the pilots who were on them for a time grew to like them. It was said they took off at 90 mph, cruised at 90 mph and landed at 90 mph and I can't fault those figures very much. Cruise was up to 125 mph if you were in a hurry.

When our Canso completed its job in Venezuela, it was assigned to an area in Nigeria north of Lagos. With co-pilot Tom Appleton, I picked up the aircraft for the ferry flight. Our first segment was into Belem, Brazil. It turned out to be a huge slum at the mouth of the Amazon and so we decided to keep on going without rest -- a big mistake. We planned to refuel at a Brazilian airforce base on an island about 300 nautical miles off the northeast tip of South America. Prior clearance had been arranged through diplomatic channels. From there we would go directly across the South Atlantic to Roberts Field in Liberia, a former military base near Monrovia, where Pan American Airlines stopped on their African service.

ALTIMETER RISING

Apart from a Bendix Doppler which went into a 'memory' (off) mode over smooth water, our basic navigation was dead reckoning, with HF, VHF and ADF all serviceable, but no one to talk with and no beacons to receive. There was little navigation on that part of the Atlantic, either at sea or in the air at that time, except a few flights to Rio de Janeiro. I don't recall any voice communication by radio on the entire trip.

Our cruise speed was between 90 and 100 knots as we had to conserve fuel. Tom and I spent the entire night looking up at the exhaust stacks of the trusty 1830-92 Pratt & Whitney radials barking away above us. We tried to get the best mixture selection that would give us the most miles per gallon. Using manual mixture control, you can lean the cruise mixture for maximum efficiency. Too lean can cause engine damage, overheating and power (speed) loss. Too rich runs the engines cooler, but wastes fuel, and we needed lots of it that far from African coast. One feature the Canso did not enjoy was accurate fuel gauges. Designed originally by a naval engineer, the PBY, called the Catalina, was basically a ship that flew. When an undercarriage was later added to make it a PBY-A (A for amphibious), it was called a Canso, but the retrofit was not a complete success. The hydraulic system never seemed adequate to handle the large, cumbersome undercarriage and the nose gear was prone to trouble. Taxiing in a strong crosswind, with a large supply of fuel in the high wing, was also a problem. The engines each put out 1,050 HPR for take off, but are brought back to 600-800 HPR for economic cruise.

When we arrived in the late evening, at Fernando de Noronha Air Force Base just off South America's northeast corner weather conditions were good and we had hoped to refuel and keep on going. However, when the Brazilian Air Force crew found gasoline for us, some of the barrels had been opened, a no-no to any bush pilot. Even some that did not appear to have been opened contained water. We always filtered fuel through a felt strainer and were alarmed to see that water kept filling the funnel. Wisely, we had declined the military offer to go to their mess and have a few drinks while they refuelled the Canso for us.

Having spent considerable time in South America and Africa, where gasoline has a high 'slippage' rate, this did not faze us too much. We had enough 'good' fuel to go back to the mainland. It is discouraging though, when you have psyched yourself up for an ocean crossing and already have 300 miles behind you, to have to go back and start over again. How much adrenaline can the body produce?

Back in Natal, on the north east coast of Brazil, we drained our tanks and filters carefully to get clear of any possible contamination, refuelled and got some food and a few hours sleep. I spent about three hours trying to send a cable to our company in Ottawa to advise them of our new flight plan. We didn't want them blowing the whistle if we didn't report in the next day from Africa on schedule but we also wanted someone to be aware if we were much too late. Air Traffic Control in that area did not cover transoceanic flights and, since we knew zilch Portuguese, any communication was difficult.

After a delay for weather in the morning, we were off by noon starting a 1,700-mile trip that took us 16 hours and 35 minutes. No coffee, no food, no sleep, no autopilot, no beacons, no VOR, no position reports, no ships or aircraft to communicate with - it was a looooooong night! We 'shot' the moon and stars with the astro compass to pass the time and to confirm our heading. There was little weather or turbulence. We could establish our drift over the water, even with the Doppler in memory so all we really had to do was wait. When you fly east, the nights are, of course, much shorter.

On this particular flight, our track crossed and nearly paralleled the equator. We were at the equator where the meridians of longitude are father apart than anywhere else. Each degree of longitude equals 60 nautical miles at the equator. With a ground speed of 100 knots, progress seemed very slow. However, the sun came up precisely where we had expected. A few hours later, we could see towering cumulo-nimbus in the distance, which is a nightly occurrence on the Gold Coast. As the clouds broke up in the morning sun we saw the beautiful African coast in the distance. Our landfall was less than 50 miles off our planned track and we were soon approaching the long runway at Roberts Field, tired but happy.

We had only had about four hours sleep in three days and were really not too rational – a bit punchy, in fact. We approached the runway straight in, reluctant to leave our cruise attitude. As we had no idea how much fuel we had left, if any, we were afraid to tilt the nose down in case we drained one or both tanks. The Canso theoretically can carry 24 hours fuel with careful handling in the loiter mode, just above stall speed. But we had flown 1,700 miles and had no idea how much was left. Finally we eased the trusty old amphibian down in a level attitude, got the gear out, and landed safely from a high approach. When we finally dipped the tanks, we found we had nearly two hours of fuel left. I wish we had known that during the last 500 miles! Incidentally, the way you check the fuel content on a Canso is to climb up on top of the wing and stick a wooden dipstick in to touch the bottom - not recommended in flight.

We spent a quiet day sleeping at the PanAm Hotel and watched the hippopotami wading in the river of the hotel garden. Food, beds, people -- flying was fun again. The cumulo nimbus cloud normally builds up in the afternoon along the Gold Coast and Ivory Coast, but our route into Lagos seemed anti climactic with lots of fuel, food and sleep. A few thunderstorms added something to look at and were better than the over-ocean boredom. The never ending drama of fishermen along the coast, some in boats farther out, most just throwing their nets into the onshore surf, passed beneath at a speed and altitude that we could enjoy. An occasional freighter offshore, being unloaded and reloaded by lighters opposite the larger towns, showed some sign of trade. Only a few African colonizers bothered to build wharves, roads, schools or hospitals. The Italians probably did more in northeast Africa than anyone to develop trade did, followed by the French in Cote d'Ivoire. The only roads the British built in most colonies were one strip from the airport to their 'club'.

An old African hand can fly down the coast of Africa at low altitude without a map and tell the political history of the terrain beneath him by the state of development. The Portuguese, great navigators but brutal colonists, put up tall phallic symbol type towers for navigation purposes. The British built their 'clubs' exclusively for themselves. The Italians planted bananas and built roads, and the French built bridges, roads and churches. Few bothered with schools or hospitals. The Americans exploited rubber in Liberia and did little to help develop the country. The Spanish built churches. Later the Canadians built missions and a network of small airports and light planes for missionary work but, of course, could never be called colonists. Since the United Nations has decreed no more colonies, these countries now have political autonomy, which means freedom to go back to tribal warfare.

For anyone interested in cruising Africa in a flying boat, a tour company offers cruises from Harare, Zimbabwe (still Salisbury, Rhodesia to me) down the Nile through its headwaters in Zimbabwe, Zambia, Malawi, Kenya, Tanzania, Uganda and the Sudan to Cairo on a Canso at low level, with landings on land and water and nights in exotic hunting camps and game parks. The pilot, at one time a colleague from Edmonton, is Capt. Ray Bernard, flying a sister ship to the Canso in this story.

CF-JJG came home the next year without incident via the North Atlantic when the contract was completed . My co-captain westbound was Peter Mitchell, a long-time friend from Ottawa, also a veteran survey pilot, fire bomber pilot and ex-airline pilot whose father was a pilot in WWII. When we came out of the Canary Islands westbound on the return ferry trip, we noted a roughness that did not smooth out as the engineer had promised. After leaving Santa Maria, the vibration got worse and after about 200 miles worrying about it we went back to land at Lajes, a U.S. military base in the Azores. New spark plugs, some shopping at the PX, and we were off again for Newfoundland, once again sleepless but with lots of fuel for the relatively short 1,500 mile trip. Westbound ferry trips are normally always against the wind and the period of darkness is much longer. You keep chasing the sun, not that the Canso gives Old Sol much of a run for the money. Of course, on this North Atlantic route, we had navigation aids, good HF communication and lots of airliners above us.

ALTIMETER RISING

I recall on arrival in St. John's, Newfoundland I was so tired I lined up on the wrong runway on final for landing, and when Pete alerted me, I said: "What the hell difference does it make? Let's get it on the ground". The tower did not object, and we got some much-needed sleep before the last leg home to Ottawa.

After crossing Newfoundland and the Gulf of St. Lawrence to P. E.I., then up the mighty St. Lawrence past Trois Rivières, north of Montreal, and into home base at Ottawa, you tend to fill up with pride at all the geographical beauty we enjoy in Canada, yet take for granted. Only after looking at the rest of the world with its slums, overcrowding, lack of water, food and sanitation, can we appreciate the beauty of our own country. And pilots are doubly fortunate. We see it all in its panoramic splendour out of the front window and so often have the enjoyment of seeing our own homeland spread out in front of us to enjoy.

ALAN MACNUTT

Incident in Colombia

The colonel was a hero and I turned Japanese

Bogota, Colombia is situated in a bowl in the centre of the country, surrounded by the Andes on all sides. It is not the easiest territory through which to fly precise survey lines. And the country poses challenges of other varieties both in the air and on the ground. This became evident in the 1960s when Spartan had a contract with Esso Colombiano and the Colombian government to conduct high-level mapping surveys. Our first winter in Colombia required a lot of PR work. To start, we had to acquire a working knowledge of the Spanish language to help alleviate the South Americans' general distrust of gringos. Don McLarty, a handsome Argentine-Canadian, was our project manager who took care of all our diplomatic problems and generally looked after us on the ground. In the air we were on our own. Although many will tell you English is the language of aviation, no one tells the air traffic controllers in most Third World countries. The best you could hope for at that time when you called to the tower in English was that they would answer in their own language. If you made a nuisance of yourself, they would send down to the coffee shop or somewhere to find someone who could understand you.

During our first contract, Rojas Pinilla was in power in Colombia and ran a form of dictatorship backed by the army. We therefore wanted to stay on the good side of the military. When we arrived, the Colombian air force pilots were very interested in our aircraft, a de Havilland Mosquito, which they had heard about but had never seen. A senior colonel in the photogrammetric section asked to be taken for a ride. Under the prevailing conditions, it was not advisable to express regret and say, 'Sorry, I can't take you for a ride. It's the insurance company, you know.'

I took the colonel flying out over the Andes around the llanos, a flat prairie-like mesa covering much of the eastern part of Colombia. We flew back over the city and over to Medellin (now famous for reasons other than its history and beauty) and started back to Bogota. One engine packed up en route, and so I feathered the propeller and we proceeded back on one engine. That was no problem in a Mosquito at about 15,000 feet. It was a safe height over the Andes in that area.

The airport at that time was Techo, before the new modern El Dorado was built. Techo, like much of the area around Bogota, sits at 6,280 feet above sea level. With the approach speed of a Mosquito at 120 knots the numbers worked out to about 175 miles an hour ground speed for approach at that altitude. The hazard increased in the emergency landing situation.

I called the tower well back and was assured I was cleared to land straight in, but on short final a DC4 taxied out on the only runway, directly in my path. Airplanes like the Mosquito fly beautifully on one engine in cruise, but when you reduce speed for landing, put the gear and some flaps down, you are committed to land unless you have lots of air beneath you. With the DC4 in our path it seemed advisable to try a go-around. The emergency manoeuvre was successful; that is to say, we didn't prang.

Like many high-performance wartime aircraft, the liquid-cooled Rolls Royce Merlins in the Mosquito heat up rather quickly when the undercarriage and flaps are lowered in flight. After beating up my good engine unmercifully in the forced go around, the coolant was boiling pretty good when I came back for the second approach. To add to my classical snowball I was having trouble making the tower understand I was coming back for another approach. On final with one feathered and the other boiling, I saw a Beech 18 sitting on the end of the only runway doing his run-up and blocking our landing. As the flight was obviously about to terminate very shortly in some fashion or other, I landed long at the side of the runway on the grass, gear selected up. Thankfully, the aircraft landed right side up. The tendency is for a twin like the Mosquito to try to roll over below a certain speed, when you have too much power on one side. Your options become limited at lower speeds. Landing on your belly is always scary, noisy, humiliating and not conducive to rapid promotion. Landing upside down, however, could curtail your career entirely.

As the colonel and I hit the fields, we collected a few concrete fence posts, surfed a few ditches, and gradually shed our speed by bulldozing terrain and shrubs. The braking power cost us some external parts, including the good propeller which 'walked away' very rapidly. One cement post was embedded in each main tank and the high octane was all about us on the ground as soon as we came to rest. I got the coupe top off over my head and, without waiting for the classic captain's exit line 'after you my dear colonel,' I abandoned ship rapidly. Emergency exit is through a top hatch, and if you anticipate a fire it is best done quickly. It is not true; however, that I put a foot on the colonel's head on my way out, as some people suggested later.

We were soon surrounded by dozens of people from the airport as well as local farmers, complete with cows. Everyone except the cows seemed to be standing in the pool of high octane fuel, smoking. When the newspaper El Tiempo carried the story with pictures next day, I could be seen trying to get the crowds to move back with their cigarettes. My urgent warnings turned out to be a wasted effort. The reporter, who seldom gets things right in aviation accidents, got my name as MAC NAK and I was identified as a visiting Japanese pilot. My Spanish obviously wasn't that good or maybe I was just a little shaken up.

Both the colonel's pride and my career emerged intact, however. The colonel invited the survey crew and me to his mess later, where he was hailed as a local hero. The company sent me home to bring down another Mosquito to complete the job and all ended well. Fully operational Mosquitoes could be purchased then for $15,000 plus $5,000 for the ferry pilot to deliver one to Ottawa from RAF surplus stores in Britain.

ALTIMETER RISING

It was not easy explaining to my wife and kids why I had to come home from South America and get another aircraft because the first one didn't work. Nor was my boss particularly pleased with my explanation that I had ground a $15,000 Wild camera into the terrain before I had even started to take pictures.

ALAN MACNUTT

R.O.N. in Santa Maria

Where the mountain pokes through the cloud

During the heyday of Survair Ltd., much of our winter aerial survey work was on the African continent. Every winter, we had to move two or three aircraft across the Atlantic to complete contracts in Africa, then bring them back in the spring. In this way, we worked around the Canadian winters and avoided the Harmattan, the fierce sandstorms that batter Central Africa. The Harmattan is started by strong seasonal winds in the equatorial regions, which stir up fine sand that rises to heights of 30,000 ft. south of the Sahal. It reduces visibility to almost nil on occasions and ruins aircraft engines. Camels and Bedouins can cope with this yearly phenomenon, but for the rest of us the Sahal is not a pleasant place to be during a sandstorm.

To implement this migration of our survey birds, we had the option of using a northerly route to cross the Atlantic. This took us via Fort Chimo in Quebec, Frobisher Bay (Iqaluit), NWT, Sondrestrom Fjord, Greenland, over the icepack to Keflavik or Reykjavik in Iceland, then down to Shannon, Ireland, or Prestwick in Scotland. From there we flew on to Gibraltar, Madrid or Lisbon to refuel for the long haul down the barren west coast of Africa to Dakar, Senegal, via Ghana, Cote d'Ivoire, Nigeria or to whatever was our final destination. Ferrying these aircraft was my job.

Starting from Ottawa, the route to Frobisher was pleasant in clear cold weather. But as the northern Canadian leg was usually done in autumn and our aircraft were not winterized, the trip could be scary in icing conditions. Only some of our aircraft had wing de-icers, and even with wing de-icers the Aero Commander flew like a drunken pig with only a moderate load of wing ice. St. Elmo's Fire, a dancing static-electric charge, was always exciting in the dry Arctic winter conditions. The props, leading edges and windscreen, were sometimes aglow on a cold lonely night on this route. Your hair would point to the windscreen like the attraction of a TV screen. St. Elmo's Fire is said not to be a hazard, but it can make you very nervous over a prolonged period. We would always carry winter safety gear and usually forget to bring it back to Canada in the spring. Once, our African inventory showed six pair of snowshoes, and no one there except the Canadians knew what they were.

After leaving Frobisher, things improved. The 500 nautical miles into Sondrestrom were easy with VHF voice communication on our alternate at Cape Dyler, a treat on most ferry flights. Best of all, in the event (not unusual) the fjord was fogged in, the USAF had crack GCA controllers to recover their aircraft based there. I have taken several co-pilots in there on a GCA approach in a storm, or at night. When the sky cleared next morning, they would realize that they had been talked down along the side of a mountain with the steep cliff less than a mile to our left. After landing at Sondre, formalities were minimal. There was a warm hangar to store the aircraft, the best in aircraft handling, great food, overnight accommodation, and weather briefing. This all cost big bucks, and had to be pre-arranged through both USAF military attachés in Ottawa and the Danish diplomats. Denmark owns Greenland.

Minimum en route altitude (MEA) was eleven thousand feet across the ice pack from Sondre to Iceland. But the flights were accommodated by reporting stations, beacons and a military airport, if required, at King Oscar Havn Island on the east coast of Greenland. Many aircraft were lost on this route in wartime ferry operations, including one flight of P38s that flew into the snow pack in formation. The War Birds, a private aviation group, have salvaged some aircraft for restoration.

An alternative to this segment was to go from Frobisher to Bluie West One, a former wartime base on the south tip of Greenland, now called Narsarsuaq. This strip was up a narrow fjord, sometimes hard to find, with no easy alternates. You had to go up the right fjord the first time, as some were too narrow to turn in. You had to land straight in when you saw the strip and take off heading out the other way towards the ocean. Accommodation was poor and few facilities existed for transients.

At Keflavik, we were handled by competent Icelanders and all the facilities for transient aircraft were good. It is a country of unique scenery, active volcanoes, and little fertile land. From a pilot's viewpoint, however, it meant good beacons and reliable radio communications. On the down side, there was lots of icing in the cloud below 10,000 feet most of the year. From Iceland on until the Mediterranean, icing was always a worry. You can get wing icing in England, Scotland or Ireland at the 700-millibar level at any time of year -- winter is just a little worse. The Bay of Biscay always seemed to hold bad weather and icing conditions from the Atlantic. The area is a hazard both for seamen and airmen.

The Scottish reputation for hospitality was never reflected in their handling of transient pilots. After they tore down the old terminal building at Prestwick, which housed the hotel, met office, refuelling office, restaurant and bank, it was a place to be avoided. On the other hand, Shannon in Ireland seemed to welcome our business. Landing at Paris or Rome was asking for grief. Even flying over France you were hassled because of our funny language - English - the language of the air! In Lisbon, the Portuguese were great record takers. Every piece of paper in the aircraft would be inspected -- log books, airworthiness certificates, radio licences, leases, and detailed records - practically a biography of each crew member, even to your mother's maiden name. Madrid was a better stop in transit, but my favourite was Gibraltar for the final refuel before Africa. In Gibraltar we landed on the RAF Base beside the famous Rock with its famous monkeys. We enjoyed hospitable accommodations, casinos and a language that was comfortable. That base has reverted to Spanish control.

After all these potential hassles on the north route, the extra 1,200 miles and six flying hours, it seemed easier to fly great circle headings from St. John's, Newfoundland, direct to Santa Maria in the Azores, 1,546 nautical miles. Then there was a short hop into beautiful Las Palmas in the Canary Islands then on into Dakar,

Senegal. This latter route was all over ocean. There were no alternate airports en route until the Canary Islands, and your first few hours out of Newfoundland eastbound, or out of the Azores westbound, you were so heavily overloaded with fuel that you would have been unable to hold altitude if you lost an engine.

Other factors were the possible loss of navigation aids on the southern route. All aircraft had two generators, but a minor electrical fault or fire could put the radios out of service, and communications and navigation aids would be lost. Although the two ocean weather ships were helpful, they were not potential landing sites and you rarely saw them. They could, however, paint you on radar and give you headings and ground speed and words -- lots of words! The ships' radio operators, bored from long periods of sea duty, would try to talk the airline crews, into getting one of the stewardesses on the radio so they could talk to a woman, even if she was five or six miles above them.

Theoretically, you could always melt your wing icing if you descended low over the warmer ocean, but that had disadvantages as well. At night it was difficult to hold an accurate height safely above the huge waves for hours on end without an autopilot or co-pilot. Ferry flights always carried the minimum legal crew. Also, to fly in the warm air in ground effect, you had to retract the long-range trailing antenna, from behind the aircraft. If too low, you risked losing it on contact with the water, which meant you, lost your HF communication with Oceanic Control. Reporting every 10 degrees of longitude was mandatory, both east and west bound, as well as at Air Traffic Control boundaries.

Our main navigation aid on this direct route between the Azores and Gander was Consol. That system was an aural aid that could be tuned on your ADF in the lowest band, about 100 to 200 Mcs (now megaherz), when you selected beat frequency oscillation (BFO) on the dial selector. Consul stations were active in Nantucket in Maine, Stravenger in Norway, Ploneis in France, and Lugo in Spain, plus one in the Florida area I could never raise. Those stations repeated a pattern of dots and dashes in a 360° circle. By counting the dots you could get a bearing from the station. The transmitter in Maine was close enough to our tail to be very helpful for a few hours on the way east from Newfoundland, as Consol has an effective range of over 1,000 miles. The system was used by the Luftwaffe in WWII to bomb England, later widely used by ships and aircraft, but now decommissioned.

We carried a drift meter to gauge our drift off the whitecaps, an astro compass and navigation tables for sun, moon and star shots at night. Generally, our biggest problem was stomach butterflies for the first few hours of each ferry trip.

One night, I was taking an Apache to Africa for a magnetometer survey and was mentally and physically exhausted when I finally reached the Azores. Pico, the tall distinctive mountain, is one of the most westerly islands and the first land you would see after hours out of Newfoundland, hopefully on the correct heading. It always looked beautiful and made me feel like Columbus must have felt 500 years ago. Pico, near the USAF Base at Horta, came into view and gradually slipped behind me as darkness fell. The high humidity reduced the 'spread' and fog patches appeared. I knew I was only hours from Santa Maria and a good rest in the old hotel made famous by early trans-ocean pilots. Jets now overfly Santa Maria, but in the early years of transoceanic flying, it was a very important port of call to refuel. It was a welcome place to R.O.N. – remain overnight.

Soon I had Santa Maria on my ADF and then on VHF communications. It was a comforting feeling after being alone over the ocean for about 12 hours, never completely sure of my exact position, or whether that port engine oil pressure is really creeping down slowly or if it is only my imagination. By now, I was on top of a solid undercast. The cloud base was low that night and the wind was high when I was cleared for my approach off the Santa Maria beacon. I still remember the frequency — 323 Kcs. Probably because of fatigue and a premature relaxation of my thinking powers, I didn't pay too much attention to the fact that when I broke out below the cloud just outside the beacon my heading was not exactly as advertised. I probably thought I had not computed the variation correctly, or was too tired to worry about minor details and descended down to minimums. At just 200 feet as I was ready to land, I realized I was not lined up on the runway lights but on the lighted road to the airport. More adrenaline was produced and I got the airplane flying again and lined up properly and landed safely. Like many pilots before me and after, I learned not to relax until the airplane is in the barn. After the brain has been exposed to an overload for long periods, it is easy to relax our concentration too soon, and lose the whole enchilada.

ALAN MACNUTT

The Yankee Dollar

Using a potpourri of Canadian resources

Canadian Aero Services of Ottawa, a geophysical survey company, started as a subsidiary of Aero Service Corp. of Philadelphia in the early 1950s but outgrew the parent company in a few years. It seemed easier for a Canadian company to get overseas contracts and the Canadian passports of our crews were acceptable in more embassies after the atomic bombs came down. Canadian survey crews had to have both FAA and Canadian licences, instrument ratings and type ratings, as we might fly aircraft registered in Canada or the USA on overseas contracts. We had to write air regulations in some countries before we could operate in that state. We often had several licences in our briefcases, like a Middle Eastern diplomat has passports.

When Spartan was fading from the scene, and before the rise of Survair Ltd., there was an excess of survey equipment. Some surplus pilots joined Kenting Aviation in Toronto, a dreaded adversary. Others joined airlines or found other work. I freelanced on contracts where I could negotiate the best deal. On one of these contracts, I joined Aero Services Corp. of Philadelphia for a DC3 magnetometer operation in South America. We were based on the northeast coast near Paramaribo, the capital of Surinam.

Surinam is a very multicultural society. Their flag of seven stars pays tribute to the seven basic groups trying to operate a homogenous democracy. The language is called Taki-Taki, a mixture of everything. The white star on the flag depicts the Dutch element and their political influence, which was still evident while I lived there. The red star in their flag depicts the Amerindian, not found much in Paramaribo; the black star represents the Africans left over from the slave trade; the brown star is for the Asian Indian; and the yellow star represents the Oriental population.

Our group leader on this project was Robert E. Lee from Texas, formerly of Pan American Airlines. We based our DC3 in Zanderij, the only major airport, where the biggest thrill socially was watching the KLM DC8 come in weekly from Amsterdam. It was at this airport, a few years later, that one of our Survair pilots lost his hydraulics while taxiing in to the ramp and came to rest with the port wing of his Canso in the coffee shop on the second level of the terminal. There was very substantial damage to the terminal and the Canso, but the pilot, Ken Fraser, never one to panic, walked out onto the wing and from the wing tip into the coffee shop and ordered a beer!

At Christmas we shut down operations. The magnetometer operator and I rented a light aircraft and flew over to Cayenne in French Guyana and visited Devil's Island, the infamous French prison. Rainfall in this area is about 110 inches per year, resulting in a mass of wet tangled vines that hide the penal colony on that ugly little island just off the coast. In Cayenne, the buildings were falling down for lack of maintenance; everything was covered with a green scum outside and mould inside from the high humidity. The people looked sad and discouraged.

During the Surinam operation, we flew over the jungle hour after hour, up to 200 hours per month, from early morning to evening with no detail to follow on the ground, just the canopy of trees 500 feet beneath. Once a day or so, we might spot some tiny clearing with a few huts along a river, with primitive boats tied up and some startled Amerindian tribe looking up terrified at the large noisy bird going overhead. The tribes in the deep jungle are very primitive, and although there is an abundance of wildlife and birds in the forest, and fish in the ocean, they still like a slice of missionary or lost pilot for dessert!

Theoretically, the DC3 will maintain height on one engine, but with an overload of fuel and survey gear, the 'good' engine would soon overheat in that high temperature and a forced landing in those tall trees would be terminal. The trees would fold up over a crashed aircraft, hiding it completely. There was no formal search and rescue and so, few downed aircraft were ever found or even searched for.

Rocky La Roche from our company was flying a Canso in this general area a few years later and lost an engine at the usual survey altitude of 500 feet. The Canso had large external pieces of survey equipment adding to the drag, and his life expectancy quickly became very limited, as the old Canso could not hold height in the tropical heat. He spotted a clearing used occasionally by a Helio Courier STOL aircraft. The strip was about 120 feet wide and 2,000 feet long. The Canso has a wingspan of 104 feet. Rocky did a sensational job of getting it safely onto the strip and all walked away. He was a very skilled survey pilot, fluent in three languages, and a respected colleague.

The winter we were in Surinam my wife came to visit and that brightened up a dull flying operation. She enjoyed swimming the local lake, called Coca Cola Creek, where the chemicals in the leaves stained the water, giving it the title. The weather was excellent and so we had no excuse for a day off. Our mechanic worked in the cool of the night and we were off again every morning for a dull, uneventful trip of 10 hours. Since then, Casinos have been introduced in Surinam and the nightlife is said to have improved.

It was snowing in Canada but Dave Rowlands, Jim Neals and Tom Appleton relaxed in the heat between flights over Africa.

When the project was completed and the data on magnetic anomalies had been sent home for completion, we went to Guyana for a photo job out of Georgetown, the capital. This entire coastal area, including Venezuela, is wet and humid and you sweat a lot. A Canadian from Langley, B.C., Tom Wilson, was operating a single engine Otter for a large mining company from an airport at McKenzie, South of Georgetown. He reported the atmosphere less humid and more pleasant away from the coast.

My first night in the hotel in Georgetown I lost my pants. When I got up in the morning and went to dress I had the strange feeling that the bureau where I had thrown my pants was moving. I sat down, shook my head and looked again - cold sober - and my pants were disappearing off the dresser. It seems that there was rubber insert in the waist of my trousers and some type of army ant had invaded the room and attacked the rubber, literally eating the waist off my slacks. I could not use them and had to send a crew member out to buy me some shorts before I could go to work. Later that week the same pests attacked our DC3 and ate the rubber bungee from the main gear, leaving it unserviceable. The early DC3s used heavy bungees for gear retraction, which were later replaced by a hydraulic ram. These ants, like the army ants in Africa, march by the millions and devastate anything edible in their path.

Survey flights over the Andes offered a variety of scenery like this volcano. The white areas in the background are salt flats.

ALAN MACNUTT

High Over the Andes

The fighters couldn't catch us

In 1963, between the heyday of Spartan and the advent of Survair, I was hired by Aero Services Corp. of Philadelphia (now Texas), to operate a P38 Lightning in Chile. The base of operations was at Antafagasta, a seaport on the northwest coast near the Bolivian border. One of their pilots had taken sick on a high altitude photo-mapping contract. Since I had experience on the airplane, with the Wild camera and in Spanish, to some degree, I was hired to replace him. The country was strange and fascinating for a pilot used to the broad east-west expanses of Canada or almost any other country with large weather patterns. Chile is very narrow, about 100 miles wide in most places and 2,200 miles long, and appears to be laid down on edge. On the west, the country slopes to the Pacific. On the eastern border, joining Argentina at the Continental Divide, is the long and majestic Andes range, much of it rising 20,000 feet, much higher than our Rockies. The climate is as variable as the land. In the far south, conditions are somewhat comparable to the Canadian Arctic, complete with penguins. The mid latitudes are rich and lush, changing from sub-tropical to tropical to desert as you go north.

In the northeast corner of Chile, in the Atacama Desert, is a large open pit copper mine then owned by the Canadian Anaconda Company. We were told that in the 27 years the weather station had been commissioned, never once had precipitation been reported. The area is surrounded by an overheated marsh that resembles a huge boiling cauldron. If walking, you must pick your footing over the ground very carefully to avoid falling into bubbling lava. The formidable barrier of the Andes between Chile and Argentina is dotted with active volcanoes. Earthquakes are a daily occurrence.

Chile is one of the few countries in my travels where reverse racism is the norm. As the history of the country taught the people to hate and fear the black-eyed Spanish Conquistadors, any foreigner with fair skin and black eyes is suspect. Only "the pure" have brown skin and obviously the Indo-Chileanos are considered the aristocrats of the revolution. People with white skin and blue eyes are a novelty and must be Gringos.

In later years, Chile suffered severely at the hands of inept politicians. The people are very proud, independent and volatile and, like the natives of California, British Columbia and Alaska, they share an affinity for the Pacific on one side and majestic mountains to the east. The comparison does not end with the geography; all these regions have experienced inept politicians.

The population of both nations are Catholic, Spanish-speaking descendants of the Conquistadors and the revolution led by O'Higgins and Simon Bolivar the Liberator, but they love to feud. To watch a game of soccer, water polo, or any sport between teams from each country is like watching an undeclared war. The players on both sides are handsome, aggressive, and fiercely proud of their country's colours. That mindset was not restricted to the playing field.

One clear day I was flying photo survey at 37,000 feet following parallel lines north and south over the Andes along the Argentine border. My cameraman said he saw two airplanes well below, going in the same direction. I had no downward visibility but he had a clear view from his bubble in the nose. As we came to the end of our photo line and turned south again to fly the parallel line, he told me the other aircraft were still below and keeping pace with us. This was a surprise as jets were rare in that area in those days and few piston aircraft could keep pace with our P38, which was probably cruising over 400 miles per hour true air speed.

It took us a while to realize that the aircraft beneath us were Argentine fighter planes. Evidently they resented us flying along their border and, in fact, turning in their national airspace. They would have liked to throw some lead at us, as we were, to them, hostile aliens. The only problem was that their B26s, also a remnant of wartime, could not climb to our altitude. They felt we were cruising around 10,000 feet above them mocking them - and you do not mock Latinos lightly, especially South American Latinos.

Our fuel eventually was starting to run low and, as we did not dare descend in that area, we flew well into Chile, nearly to the Pacific, before we felt safe to start our descent, even over friendly territory. When we returned to our base in Antafagasta we were hailed as heroes by the Chilean Air Force. They had been advised through diplomatic channels of the 'attack' that their planes were alleged to have made over Argentina. We were wined and dined in their mess and, of course, our egos enjoyed being stroked as the brave Gringos who had taken on the hostile Argentine Air Force and found them wanting.

The Lockheed P38 saw much action in WWII. After the war, survey companies pioneered by Spartan cut off the nose section containing most of the armament, put on a clear plastic nose, something like the bubble on a helicopter, to house a camera operator. Extra oxygen and long range tanks were added, and presto, it was a photo aircraft. It is interesting that this aircraft was designed in 1937 by a young engineer who was earning $25 weekly at the time and just out of college. He designed a winner that made air war history in Europe, the Far East and Africa.

To fly the P38 was a pilot's dream. It was the first of its type to have hydraulically powered ailerons, a super comfortable cockpit, and many other features to please a pilot. In the creature comfort department it was unlike the British fighters of the era. It had counter rotating propellers, one turning clockwise and the other counter-clockwise. This cancelled out any swing on the take-off roll. Unfortunately it was a maintenance nightmare, especially when used above the 30,000-foot range. The induction system would break down and frequently an engine would lose its liquid coolant. For economic reasons De Havilland Mosquitoes later replaced P38s in Spartan. That fighter-bomber was noted for its practical, efficient, no nonsense Rolls Royce Merlin engines and rugged dependability. The Mossie lacked the niceties of the P38 but it was always ready to go to work.

ALTIMETER RISING

The North Sea Find

The shuttle between Denmark and England

Long before the British and Scandinavians developed North Sea oil, Survair/Canadian Aero spent an interesting year doing a survey across that area. Our platform was a DC3 equipped with long range tanks, Decca and Doppler for navigation equipment, a radar altimeter, and state-of-the-art survey equipment. We were based in two countries, had full time hotel rooms in each and worked like beavers on that job. After lots of compilation by geophysicist, Dr. Art Rattew, and other specialists, the results became known as a major input to the British economy and a new source of oil for Europe.

The crew included Norm Spraggs, a crack engineer, an eligible handsome bachelor (at that time) and a great friend over the years; Garland Erickson, a charming Danish-Canadian rogue from Laurentian Air Service, my co-pilot; Donald Davidson, a British Canadian electrical engineer, who was in charge of the survey gear. Our transits extended from the English coastline to the Danish coastline. With the daylight available we could do three complete lines daily and have time for refuelling before darkness.

Our two bases were at Manchester and Copenhagen. Donald was from Manchester and Garland grew up in Copenhagen and so this international 'leg up' made their value to the crew two-fold. With care and attention, Norm and I, both Canadians, could talk to the ground people in Manchester, but in Denmark, Garland was our interpreter. People who have travelled in Scandinavia know that English is the second language and most speak it well. But the rest of us did not know this at the time and Garland insisted he should interpret for us on all matters in Denmark or we would be ripped off unmercifully. The only one who ripped us off was Garland himself. For instance, when we landed there first to set up the long term operation, he insisted we did not show ourselves until he had made all the arrangements for rental car, hotels, ground handling, refuelling, etc. He said he could get much-reduced rates through his linguistic attributes, friends, and general savvy. He finally came back to collect us driving a big Mercedes that he stated the rental contract only permitted him to drive. And this was only the first scam. When he checked us into hotels he always got the suite. We got what was cheap.

Room reservations and logistics were not a simple matter in Britain where you must 'book'. No one took reservations on the telephone. It was easier and cheaper, therefore, to keep our rooms at each end even though we only used them every second night -- one night in Manchester, the next in Copenhagen. We were not around either hotel much, and so the staff in Manchester did not know us well. One night Norm Spraggs came back to the hotel (he was probably out working on the airplane until 3 a.m.) and found the hotel main door locked. The night watchman refused to let him into the hotel. Not one to give in easily, he began to throw gravel at the windows, trying to wake some of the crew to let him in. But he only succeeded in alarming other guests who seemed a humourless lot. Worst of all his sins was that he had a bad colonial accent, which most natives thought very improper - especially at 3 a.m. Next morning's flight was delayed while we changed hotels.

I had several humbling lessons on this operation, something pilots need often. One of them involved instrument flying. Like any other professional flyer, I considered myself a hotshot instrument pilot when this operation started. But I soon found that I was a rank amateur compared to the European airline pilots. Coming back to Copenhagen the first time, the fog seemed routine to a veteran like me. Kastrup airport is at the edge of the cool water and fog was the norm, just as it had been at Churchill. All the aircraft ahead of us were getting in -- no problems. The traffic seemed normal. But when we made our first approach in the fog to minimum legal height we saw absolutely nothing. We had to go around again in heavy IFR traffic. Eventually, we got another approach clearance but missed again! Airlines from every nation in Europe were taking off and landing in an orderly fashion, yet this confused sounding DC3 pilot with the Canadian accent had missed twice. We had become a nuisance. On the third approach the controller offered to give us assistance and we permitted a GCA operator to talk us down. We saw the lights, landed, and in a state of shock taxied to the terminal in the heavy fog.

This dense fog, heavy traffic, and crowded airspace was a daily occurrence for European airline pilots in the 1970s. In Canada, landing in dense fog would only happen on either coast and then only occasionally. Blowing snow on the prairies was a cinch compared to the weather in Northern Europe in winter. On the English side, with which we were more familiar, the visibility at the airport was terrible in rain or shine. Although it was somewhat better than wartime, the visibility in Manchester on a clear day was two to three miles in smog. When the spread narrowed, it got much worse.

In a typical day's operation we would leave our hotel in Manchester early in the morning, take off on a 10-hour trip, climb out IFR on airways to the U.K. coast, cancel IFR and descend east of the coastline to 500 feet over the water. Establishing our height by radar altimeter and position by Decca backed up with Doppler, we would start our line as close to the coast of Scotland as we dared, then turn on our transit direct to the Danish coast nearly three hours away. Often we would do whole crossings without ever seeing the North Sea below us. Several days we did not see ground or water from the time our gear came up in the morning until we were on short final on the ILS in the evening. After three lines, turning as close to shore as possible, we would request an IFR clearance to Copenhagen, climb up and do an ILS back into Kastrup.

Decca is a system of navigation that is popular in Europe and eastern Canada mainly for ships. It is effective for aircraft within a limited area, but we had to use two different chains -- the British and the German -- to get coverage for such a large area. The complication came when midway in the North Sea we would lose our signal as we flew away from the transmitter and it took some time to pick up and prove the other chain. Invariably when we did, the new chain showed us off line. Which was right? It must have driven the compilation crew crazy, because unlike most such surveys where the tracking camera will solve such discrepancies, we were over water and usually had an undercast in any case.

In Peter Newman's book, The Western Establishment, Volume 2 (circa 1981), he stated that a Calgary oil drilling company had sunk 200 dry holes in their North Sea lease before finding oil. Then they found 1.2 billion barrels, equal to one quarter of Canada's supply. Could our discrepancies in survey have caused this?

Once a geological anomaly has been detected, its precise location must be mapped and eventually recovered and exploited. Our Bendix Doppler would give us some continuity as we changed Decca stations, but the accuracy of that transitional area was questionable. Loran C, GPS and Inertial Navigation, all developed since those days, would have precluded this problem. Evidently somebody recovered the right anomaly, and the rest is history.

This reminds me of the story of a bush pilot flying in the Labrador area that noted that his compass went crazy one day as he flew low over a certain hill on his sked. As pilots are men of sound ethics, he is alleged to have quit his job quietly, bought an aircraft and went back a few weeks later to find the exact spot where he was sure there must be a large burden of magnetite or iron ore. He never found the correct hill, but in later years Wabush, Labrador City and other mines proved his ideas were sound. Few survey crews, to my knowledge, parlayed their inside knowledge of survey findings into money on the stock market. Whether because of company loyalty, fear of lawsuits, or lack of cash, I don't know. Better to assume aerial survey crews are a superior race of people, untouched by greed or other base instincts common to lesser men! Not all survey jobs were as rewarding as this job over the North Sea. Only a few years later, giant oil rigs were in the area we had flown, pumping up liquid gold from the floor of that turbulent body of water.

Aerobatics in Morocco

Looking up at the Sahara

Canadian Aero Services Ltd. was a geophysical exploration company based in Ottawa in the post war years. Its prime objective was to carry out magnetometer surveys over much of Canada and many foreign countries. The company was founded by a brash young man named Tom O'Malley who arrived in Canada one day on the train with a magnetometer from his parent company, Aero Service Corp. of Philadelphia. He built Canadian Aero into a large survey group, which gained international respect and eventually dwarfed the parent company. Canadian Aero did not buy aircraft, at least not initially, for two reasons. Canadian laws precluded Americans owning a major share in Canadian aircraft. Secondly, Tom knew that there was no money in operating airplanes when you could rent them cheaply. Spartan Air Services, Bradley Air Services and Survair Ltd. were all dancing in line to provide rental aircraft. For 20 years pilots could expect their pay cheques to be signed by any of those three companies although we flew the same aircraft on the same jobs.

Canadian Aero recruited many electronic engineers, data compilers and cartographers; one of them being a young electrical engineer from Edmonton named Dwayne Olson. The company moved him to Ottawa where he was instrumental in designing and operating advanced survey systems. He refined the means of finding ore bodies using scintillometer, magnetometer, and electromagnetic devices. Above all, he worked on the tracking cameras and navigational devices essential for survey. It is of no value to discover a gigantic body of iron ore or uranium somewhere in the

Canadian Shield, North Sea or in a tropical jungle, if you cannot pinpoint that exact spot later. It's called 'recovery'.

In the late 1960s, Dwayne and I were assigned a project in Morocco. I worked for Survair flying the aircraft, and he operated the survey equipment for Canadian Aero. We were using an Aero Commander 680E - the 'E' meaning extended wingtips (an extra tank inside). As our customer was Spanish, we had a young engineer with us named Señor Mateo Arroyo from Madrid. Casablanca was the nearest major airport to the survey area, but we used Csar es Souke in the interior for a few weeks. Our survey area was immediately south of the Haut Atlas range and just north of the Sahal - a desert area with little water, camels, Bedouins and sand. Temperatures were in the high 30s and 40s during the day but when the sun went down and the temperature dropped to 15 to 20 degrees C. We felt the 'cold' severely. Our bodies soon adapted to any temperature – it was the change that hurt.

Flying survey lines, hour after hour, is a boring business, especially in turbulence over featureless terrain. Our transit lines on this job were parallel to the mountains and north of the Sahal, just a few hundred miles south of Casablanca and Marrakesh. In aerial survey you fly a series of parallel lines covering the area. Then, by means of a central line across the area diagonally, the whole block is tied together for compilation. This is done photographically by means of a strip camera, which runs continuously. The magnetometer tape shows potential ore bodies as anomalies which are recorded for study in the lab by a geophysicist.

Initially magnetometers were towed on a cable behind the aircraft in a bomb-like capsule, about six feet long and a foot in diameter. Great care had to be taken by the crew when hill climbing to avoid snagging it in the trees. Our normal flight pattern called for us to keep the 'bird' 500 feet over the ground on most jobs. In mountainous terrain that was very demanding on the pilot. Wide altitude variations would distort the data and make compilation more difficult. A radar altimeter was used to keep us honest. This device produced a tape showing our altitude for the duration of the flight.

ALTIMETER RISING

You have to decide - just before the airplane starts to shake, warning of an impending stall - which way to turn to clear the trees as you turn back in the opposite direction 180° to get more speed and altitude. Then you turn back into the hill at exactly the same spot you broke off so that the camera and magnetometer can pick up and continue farther along the line up the hill. In mountains you may have to do this several times on some lines to clear the hilltops. In turbulence it can be very demanding as you are working near the stall, near the ground. Even in moderate turbulence there is the possibility of severe downdrafts on the leeward side of hills.

As survey evolved, the bird was attached to the tail of the aircraft in a phallic-looking device called the 'stinger'. The aircraft equipped like this became known as male planes. Many other combinations evolved, and are still evolving, to tap the mysteries of Mother Earth.

A normal 'mag' survey day could start shortly after 6 a.m. in the tropics, fly until noon, refuel - lunch if possible - then fly in the heat of the day until sunset at 6 p.m. Tropical twilights are very brief and when the sun goes down it's GONE - none of this slanting across the horizon that we see in Canada. This form of survey, unlike aerial photography, has changed little in the last 20 years. The addition of Loran and GPS as navigation aids replacing the old eyeball method has resolved some control problems. Just weeks before writing this, I met two old friends, Dave Fenwick and Serge Malle, flying survey in Northern Alberta using the same time-tested technology. All three agreed that old survey pilots don't get older, only better. They were flying a twin Beech Queenair. It had the same stinger on the tail that we had used thirty years earlier. The aircraft was greatly improved as was the data produced for their customer's geological map. The search for mineral wealth never ends.

Normal turbulence over hot terrain in mid afternoon is a fact of survey. I was always strapped in, but Dwayne needed to move about to check his equipment and did not wear a harness. There are always a lot of loose parts; paper, ink, film, tubes, tools, etc., drifting about in the aircraft. They are hard to secure, being needed from time to time in flight.

On one occasion we hit a severe vertical current that threw us right over on our back. We were not mentally prepared for any aerobatics that day. To quote the time honoured pilot's phrase "There we were at 500 feet on our back . . ." etc. The law of gravity responded more quickly than the pilot in command and immediately we had bottles of printers ink all over the cabin, film boxes, camera crates, spare tubes, a Texas Instruments' printer, tools, cushions, and maybe even some fertilizer on the ceiling. The nose was dropping and nary an oasis in sight. When I finally reacted to this unexpected drama, I rolled the Commander around the rest of the way. It wasn't an airshow type manoeuvre, I'm afraid, but it got us back in control again. The inside of our airplane was a complete shambles, but at least the instrument lettering was upright and all the junk, plus years of accumulated litter, was back on the floor where it belonged.

Dwayne and I met recently and discussed this trip and, as I recall, his version goes like this:

"We were cruising along, minding our own business, bored out of our skulls, stunned by the heat and hours of noisy engines. Suddenly, I was on my back on the roof of the airplane. Two large glass (prior to plastics) water bottles, about five gallons in size, were coming up to the ceiling to join me. (We always carry lots of water on board in the tropics.) When everything loose in the airplane was on the ceiling with me, you rolled the aircraft right side up and it all came down on top of me again as I hit the floor first."

We returned to Casablanca promptly - took the day off and, with our Spanish colleague, disproved the myth that Arab countries do not have strong spirits available. Casablanca has many charms. The architecture is superb. French and Arabic are the languages, and the food, hotels and casinos were world class. The souks (markets) and casbahs (back alleys) were of course not good places for infidels like us to be caught alone at night.

Operations in foreign countries always required a lot of careful adjustment to local customs. But it was somewhat disconcerting to be trapped in an etiquette warp in our own country. On one project from a base at Tofino on Vancouver Island, Dwayne and I removed everything from the aircraft that could come out and still let the airplane fly. Our overnight kit (as we would fly till dark then land at the nearest airport) consisted of one tube of toothpaste and two toothbrushes, no razors, clean clothes or even emergency supplies. In order to climb up and down the coastal range in a light aircraft, we had to minimize our gross weight. One night we had to stay in Victoria. We had little cash and credit cards were a novelty. The only place that would take our plastic was the Grand Dowager of Canadian hotels -- the Empress. The maitre d', whose normal smarmy demeanour was reserved for serving English tea to rich tourists, was very unimpressed with our dirty clothes. Our unshaven, unkempt appearance, and BO from sweating in the hills all day forecast the dubious chance of a tip. Every job has its problems.

It didn't carry the mail but this DC3 was called a 'Male' plane because of the long stinger it carried on its tail. The device sensed the presence of mineral anomalies below the earth's surface.

ALAN MACNUTT

The DOT Years

Riding the jump seat

I was living in Ottawa in 1968 and had just been relieved of my responsibilities as Operations Manager for Survair Ltd. I was feeling sorry for myself. But with the same kind of luck that got me rescued by Canadair, an opening came up for an Air Carrier Inspector in the Department of Transport. That was considered a plum job at the time. I did not have the required airline background, but I was known to the DOT brass. My authority to do proficiency check flights and instrument rating renewals with several survey companies was a big plus. But my greatest asset was a friend in Jimmy Wells. In addition to being a fellow Prince Edward Islander and a lobby lawyer in Ottawa for the aviation community, he was a major influence in several companies for which I had worked over the years. His intervention on my behalf gave me instant credibility, and I got the job.

The Department of Transport is the federal body that sets standards of safety for licensing of pilots, engineers, aircraft, airports, airways, and air carriers. The inspectors visited all carriers operating large aircraft and the smaller ones licensed to operate under Instrument Flying Rules. We would monitor overall crew proficiency. That meant doing recurrent instrument proficiency rides, observing standards, checking their manuals and equipment and reporting any irregularities. As a general aviation veteran, I believe I had some helpful ideas as to how the smaller carriers could overcome some of these irregularities without incurring bureaucratic wrath or great expense and still meet the required standards of safety. The DOT has been re-invented as Transport Canada and its air navigation services hived off as Nav Canada.

The six years in Ottawa in this position were happy years. I was trained on a number of aircraft types that I never would have had a chance to fly otherwise. I also got great satisfaction visiting progressive companies and finding the competent supervisors providing disciplined pilot training on a regular basis. Such was not always the case. The travel especially was interesting. Regular inspections took me to Moscow, Tokyo, Australia, all over North and much of South America, and to Europe at least once a month.

In addition to our duties in flight, we were expected to report on ground installations, such as dispatch facilities, passenger handling, fire hazards and safety related problems. On such an inspection to Moscow, Captain Steve Albulet of Air Canada was in command. On take-off from Copenhagen to Moscow when the first officer reported "positive climb" (a mandatory call by the first officer). The captain knows that the aircraft is in the climb safely can set his new attitude. He then called for "gear up," which is also normal. But the call came out as a roar. The strength of Steve's voice over the noise of the four jet engines on the DC8 sounded alarming. I was seated immediately behind him on the jump seat. Like me, Steve grew up with noisy aircraft with poor if any headsets. His hearing damage had made him holler his commands and I was unprepared for such noise on the flight deck. Shouting and loud discussions or undue emotions are unusual on a flight deck. Crews sit quietly and work in an atmosphere of calm and mutual respect. Later, I learned he was called 'Shouting Steve.' He was a colourful character and a very competent pilot. After his retirement, he joined his son in operating a small charter airline out of Fort Simpson in the Northwest Territories.

Halfway to Moscow, before we could leave Western European airspace, we had to call Russian Control and establish contact and gain permission to enter Russian airspace. If no contact were made, we would have to return to Denmark. The air corridor into Russia was very narrow, communications were very 'iffy' and we would not have been surprised to see Russian fighter jets coming in alongside for a positive identification. On arrival at Moscow we passed a large tower near the airport which seemed like a serious hazard during bad weather. After a series of communications that were difficult to understand, we were given permission to land.

A very large number of Russian aircraft were all over the airport and I was surprised to learn that Aeroflot, the Russian airline, also owned and operated all the military aircraft. It was the largest airline in the world at that time. This was an unusual concept to me, believing the airlines and air forces of a country were vastly different and had little in common except that they both flew aircraft. The Moscow Airport, at that time, was a mess. I was not allowed to visit the tower. The bathrooms were all blocked, the tarmac, with a lot of litter blowing around, posed a hazard if ingested by jet engines. The terminal was a worse mess. People and police milled about like sheep.

The biggest surprise to me as a victim of our political propaganda at home, was the freedom I had to circulate. I went downtown, walked about Red Square with a camera, and was never questioned or bothered wherever I went. Intourist, the Russian travel information agency, was helpful in finding accommodation and I was free to push and shove and be shoved in Gum, the big department store. There was little of interest to buy, but a small shop where they accepted only 'hard' currency, sold me some souvenirs for my family.

Both men and women worked in the streets, building and repairing roads and buildings. The workers all looked the same - huge, square, heavily dressed in thick, long coats, hats down about their ears, expressionless, flat, dull faces that never smiled. Men and women looked the same and it was hard to tell the difference.

The Moscow visit made me realize what a cushy job I had. Working conditions in the Canadian government are far above the level a bush pilot would expect. Things like overtime, pensions, regular holidays, the comforts of home or top notch hotels, puts government employees in a different status from bush pilots. The perks are so generous that people who have spent more than a few years in government service are very reluctant to leave as they "can't afford to quit." The deciding factor for me was the transfer of inspectors to various regions of Canada. I had the bad luck to pull the Edmonton straw from the hat. After several years of route checking DC-8 crews to Europe and Hawaii out of my home base at Ottawa, I was now based where all the carriers for whom I was responsible operated into either the oil patch or the Arctic. I had paid my Arctic dues, and I had been spoiled working with professional airlines like Canadian Pacific, Wardair, Nordair, and Air Canada. I was not similarly impressed with the standards of some Alberta carriers at that time. After two cold and miserable winters based in Edmonton, I started looking for a job in 'real life', preferably where I could do the flying where the snow didn't blow in winter and the sand in summer.

I had no regrets leaving the DOT. There were many excellent people there during my term. People like Max Campbell, Rollie Langlois, Merv Flemming, Gerry McInnes, Ed Blair and Eric Bendall, all had excellent airline backgrounds and were intent on doing a good job. Charlie Fortin joined after retiring from Quebecair and Ed Knox after retirement from Canadian Pacific. Both were also veterans who had paid their dues and were entitled to and enjoyed the respect of the industry. There were others whose main objective was to keep from making any decisions until their pensions were secure.

We were often treated like academics on the flight deck during the route checks. Some flight crews, especially in Air Canada, gave the impression that we were comparable to instructors. "If you can't do it, you instruct others" or "If you can't get a job in an airline, you criticize those who can." Being, at that time, a government appendage they considered us powerless to discipline them and incompetent to advise them.

Another downer in the 1970s came when 500 experienced pilots were discharged from the RCAF and many joined the department. They were nice guys and good pilots, but knew little of civil aviation. They had never rolled a gas barrel in the north, and it took them a few years to learn that civil aviation accountability was quite different from the peacetime RCAF. Our departmental handling of the smaller carriers deteriorated. Suddenly expense claims and pension benefits seemed more important to some inspectors than the plight of small airlines struggling to meet a payroll.

These years I spent looking at civil aviation from the government side were a valuable learning experience for me. I had spent 22 years in general aviation, some of it in supervisory positions. Now I had sat on the other side of the desk and maintained much sympathy for the many small carriers who were struggling to remain solvent in a tough business. They could not survive too much static from bureaucrats who were "trying to help them". On the other hand, it was a shock to see how some of the small regional charter companies operated with inexperienced crews, unserviceable, overloaded aircraft, and often in weather conditions experienced pilots would have avoided. Often the chief pilot or operations manager would be a well-meaning man who wanted to observe the many safety regulations in their operations manual. Economic pressures from the owners, however, often did not permit the hiring of experienced crews or the training of the pilots and staff already on the payroll. On our base inspections we tried to ensure that they had at least one strong supervisory pilot who had company authority to set a safety standard, if not indeed meet the fine print of the regulations.

In retrospect I can safely say that although in later years DOT did little to encourage general aviation, they did not intentionally try to hinder the smaller companies. Many smaller air carriers, however, and at least one big one (Wardair), were actually dragged down by the absurd rulings by the Air Transport Committee, the body in charge of licensing air carriers and route approvals.

In all, I was glad to get back to civilian life with all its insecurities and pitfalls, after eight years in government service. I missed all that overseas travel that I'd enjoyed for many years in survey and during my early years with DOT. After being back in civil aviation for several years, I accepted a short-term contract with the International Civil Aviation Organization, a branch of the United Nations, as Assistant to the Director of Civil Aviation in Botswana, Southern Africa. It was nice getting back to Africa again. I enjoyed visiting Johannesburg on weekends and trips to Swaziland, Cape Town,

Durban and Harare. Victoria Falls was still beautiful to visit but a political hotspot.

Part of my responsibility in Botswana was to help put the Botswanans into flying, technical and administrative positions in aviation. As it was clear that this would not happen quickly, I asked that my contract not be renewed. My job there was much like going back into the DOT. ICAO, like its parent the United Nations, the League of Nations before it, and communism, are all great theories but get lost in the administrative process. ICAO, in my experience, thrives on meetings, none of which are very productive. There are usually far too many delegates with far too many hobby horses. All must speak at length in democratic fashion. Then the meeting breaks for photo opportunities, lunch or cocktails. Meanwhile, a few underpaid clerks are putting the draft together using language that has been agreed on weeks before and touches on each point mentioned without making any decisions. The convention then is hailed by the senior delegates as a great breakthrough for their personal opinions, whereas nothing has been accomplished.

Walt McLeish, another former Mosquito pilot and our Director of Civil Aviation at the time, appointed me as a Canadian delegate to a series of conventions to resolve the problems of hijacking known as the ICAO Seventeenth Amendment. My job was to provide pilot input to our Canadian delegation. I never learned why I was selected for this job as no one has ever accused me of being diplomatic and I had no legal training. It gave me a close look at how international agreements are reached. As many as 150 countries with delegates speaking many languages, sit down to seek support for their own political ideas.

The first conference was held in The Hague, in a huge meeting hall not far from the famed International Hall of Justice. We all listened for a month to the colourful and meaningless prose of the Spanish, the blunt negativism of the USSR, and the French delegates explain how logical they were, and so on for days of opening positions and flag waving. After several hundred speeches and as many photo appointments and cocktail parties, the meeting was adjourned a few months later to London to the Royal Albert Hall, where the whole process was repeated *ad infinitum* with great pomposity and pageantry. But little was being done to cut the threat of hijackers. Finally, months later, we all met again at ICAO headquarters for one more month of negotiations in Montreal and an agreement was signed, which has done little to protect civil aviation.

The first item, for example, was to fence all airports. In Europe it makes sense to keep undesirables off the airport premises in that crowded environment, but on the Canadian prairies, few but the gophers were affected. It confirmed however, in my mind, that Canada is a very small player on the diplomatic

Diplomacy was more frustrating than the most hazardous flights for Al (right) Here he confers with A.J. Fonteign during a hijacking conference in The Hague.

scene. We are very young and sometimes naive in international affairs. We seem too keen to rush to the assistance of our southern neighbour when they need an extra vote - regardless of the consequences.

Much of the training I received in DOT was never used. I was given the ground school, simulator and flight training and received my licence endorsement on several airplanes. I flew the Jetstar, a beautiful high performance four-engine jet, which was used by the government's executive pilots to ferry VIPs about Canada. This is the type of aircraft Howard Hughes used to train his crews to fly the Boeing 727. It was efficient, cheaper to buy and operate. Hughes had his engineers tape the two middle thrust levers together so that it handled like a 727, with three 'throttles' instead of four. The power settings, mach numbers and speeds, were almost identical.

ALTIMETER RISING

I was also trained on and flew the Viscount, Hawker Siddley 748 and the Fairchild F27 and F227. Excellent training and good experience, but like the French I was sent to study, regrettably not used a great deal.

One DOT inspection renewed old friendships at Atlas Aviation, Resolute Bay, NWT. The owner was Weldy Phipps who pioneered many High Arctic routes. Resolute Airport is now called Phipps Field.

ALAN MACNUTT

Flying with the Tsetse Flies

It doesn't get more dangerous that this

Botswana (formerly Bechuanaland) achieved its independence from Britain in 1966 and has prospered as a well-run democracy. In the northern part of the country, Maun is the only town of any size. It is the centre of a large tourist industry catering to big game hunters who fly into the many game camps by small charter aircraft. The hunters spend lots of money to hunt in the Okavengo Delta. Botswana has no major rivers, but when the rains come in the north the Zambezi seasonally floods a large flat plain. Rain gives the desert lots of moisture and a new life and grazing for animals for several months. Rain is so scarce and so important to Botswana that the word for money and water is the same -- *pula*.

This is also the best game country, a feature that keep farmers and 'white hunters' in constant dispute over the clearing of land for grazing versus keeping the swamps for wildlife. A key factor in this debate is the tsetse fly, which can paralyse cattle and devastate a herd, but which has little effect on the wild animals. It does, however, have a catastrophic effect on humans.

During my posting in Botswana as assistant to the Director of Civil Aviation the main objective was to train locals for the takeover of their own government aviation services and their airline. My wife joined me in a rented home in Gabarone, the capital. She was robbed the day she arrived and twice during the subsequent year. She then became more street-wise. Despite those incidents we both had a pleasant, if frustrating, sojourn in Botswana.

One of my responsibilities was to fly with all the carriers to assess their safety standards, efficiency, pilot training and adherence to air regulations. One carrier that caught my interest was a Mozambique company that operated a fleet of Shrike Commanders. These light twin-engine American-built aircraft were equipped to spray the dreaded tsetse fly. This was done with a large on-board tank of toxic chemical pumped out to lateral booms and sprayed through nozzles in a swath span of about 150 feet. Electronic navigation kept them on line as they flew back and forth on parallel traverses.

A normal spray operation you say? Not so! This spraying was done at night when the insects were on the ground. Using extremely bright search lights in each wing tip, the pilots maintained a height of about 20 to 40 feet above the treetops, which fortunately all seemed about the same canopy height.

The biggest hazard seemed to be that the trees were filled with birds, big white birds (that looked like seagulls to me) called tick birds. When startled from their roost at night by an approaching airplane and two powerful lights, the birds would launch straight up from the trees into the flight path. This did not happen occasionally, but constantly over the entire flight every night. To minimize the hazard, the company had chosen the Shrike for its sharply-sloped windscreen and a sleek nose. A metal guard was placed in front of the windscreen to keep the crazed birds from entering the cockpit. The pilot's reaction time amazed me as he avoided those plump birds which could easily punch out a windshield and cause chaos at that operating altitude.

The human toll

In spite of efforts to eradicate the tsetse fly in Africa, its bite today threatens 55 million people in 36 countries with the dreaded sleeping sickness. The problem has been documented since the 14th century. The World Health organization estimates that 300,000 case of sleeping sickness would be diagnosed if wars and social unrest in general did not prevent adequate testing. Each year, 25,000 cases are added.

Death in a most slow and agonizing form is inevitable if the disease is left untreated. It has been described by UN medical authorities as an epidemic of catastrophic proportions.

When the birds come up out of the trees about a quarter mile in front of the aircraft, they would head straight for the lights. Then, just as an impact seemed inevitable, they would dive. What happens when the aircraft passes inches above them and they hit the slipstream after being blinded by the glare, is hard to imagine.

The pilot explained to me, in a mixture of Portuguese and English, that when birds encounter an aircraft in their flight path, they always dive. So, a pilot faced with such a hazard, should program himself to always raise the nose. This also puts the windscreen on a flatter, more nearly horizontal surface, which if struck might deflect the bird off harmlessly.

As a 'bomber' pilot I was paid to drop chemicals on forest fires for about 18 years, from a safe altitude, with a powerful aircraft, in broad daylight. A major safety feature was the supervision of a Bird Dog pilot flying ahead of me. And I frequently am told I had a hazardous occupation!

To spray jungles at night, hour after hour from very low level, with toxic chemicals all around you, and huge birds springing up in your face like the 'wave' at a baseball game, takes more intestinal fortitude than I possess. I felt my responsibility to do an evaluation on this carrier had been completed after my first flight as observer.

I knew a pilot in Nairobi - Jack Pierce - who flew similar missions at dusk in Kenya to bomb large trees where a species of birds nested at night. His mission was to drop toxic chemicals to kill them. These birds devoured the corn with the appetite of vultures. Like the locusts, their swarms block out the sun, and reduce everything where they feed to a soggy, lifeless, stinking wasteland.

During my posting in Botswana, I met several Canadian pilot/ missionaries flying about their jobs and one Canadian-registered commercial aircraft; what else but an old Canso that I had flown years before on aerial survey for De Beers the diamond company. The captain was Serge Malle, a Canadian veteran survey pilot.

When my contract was completed with United Nations in Botswana in 1981, we elected to come home 'the other way around'. It is only slightly farther from Johannesburg, to Vancouver via the Orient than via

Europe and the Atlantic to North America. Our first stop was the Seychelles in the Indian Ocean, a beautiful group of islands about a thousand miles east of Kenya. Now a tourist resort with an international airport and several smaller airports on the outer islands, it was inaccessible by air until the 1960s. During my years in survey, our company had a contract to map the islands. This was done with a Noorduyn Norseman on floats with a hole cut in the belly for the camera to shoot down between the pontoons.

Norm Spraggs was in charge of the project and I'm sure he hated to see the job end. The climate is beautiful, the air always smells of spices, and the natives are friendly. Another pilot friend, Ed Kozystko attempted to organize a flying boat service using a Canso amphibian between Kenya and the Seychelles in the 1960s, but was bogged down by the inevitable bureaucracy.

An A26 bomber drops a salvo of retardant on a western wildfire.

ALAN MACNUTT

Conair Aviation

Bombing the wildfires of the west

When I quit the Department of Transport in 1975 to go back to private enterprise, my first job was with a West Coast company called Conair Aviation Ltd. The company had started as an offshoot of Skyways of Langley. It operated Stearman aircraft on agricultural crop spraying as well as forestry contracts to control the spruce budworm. When I joined Conair that January, the Stearman had been replaced by Grumman TBM Avengers, Douglas A26s, and Douglas DC6s. The Avengers had been navy submarine bombers, the A26s U.S. military fighter/bombers, and the DC6s passenger/cargo aircraft. All had been turned into forest fire bombers. In addition, there was a wide selection of light aircraft used as spotters and bird dogs. There was a Harvard used as a toy. The basic bird dog aircraft used to direct the DC6s was the Ted Smith Aerostar, a light high-performance twin.

One day I had to take the Aerostar into Revelstoke in the mountains of B.C. to pick up a passenger. The weather was beautiful, the mountains spectacular as always, and the trip a pleasure. However, on arrival in the early morning, the airport was fogged in. The sun had not yet hit the Columbia River valley.

I did a low pass but there were no approach aids and so I pulled up and waited a while for the fog to lift. That normally happens quickly when the sun gets up at a higher angle, but on the third pass I still couldn't find the runway. Often in these circumstances when you fly over the airport at 1,000 feet or so, you can see the runway clearly through the fog looking straight down. But on approach you can't see the button of the runway until you have passed it. Usually it is on one side or the other at about 100 miles per hour or more. Then it's too late to land. On my fourth pass I found the button at the north end and landed safely. Immediately, I ran into a bank of fog at the far end that was so thick I couldn't find my way to the office of the local feeder airline.

Just before this event, during my stint as an Air Carrier Inspector, I had concerns about the carrier who served this area and I was constantly harping at them for breaking regulations, in particular, bad weather operations. When I finally found the terminal in the fog, the boss of the feeder airline walked out to the aircraft to find out what crazy pilot was flying in that weather. We got quite a surprise to see each other! But we share a friendly laugh over the coincidence.

'Toby' Tobiason, a long-time DOT inspector, once told me, 'you don't know who your friends in aviation really are until you leave the department.' Toby and I both left the Department of Transport at the same time to join Conair. He stayed a while then went back to DOT. However, I was frustrated with the Department of Transport and I hated living in Edmonton. I wanted to be back in Civil Aviation, to do more actual flying and to be able to expressed ideas and suggestions without worrying that it would upset the government's political status quo.

After the restricted amount of actual pole time (hands on flying) that I had been doing in government, checking out in all these new aircraft was a pilot's dream. The company, at this time, had a very bad accident record and as chief pilot I undertook to try and turn that statistic around. My job was to write operations manuals to reflect how our company would comply with the various Air Regulations and Air Navigation Orders that were applicable to our fleet. Since the company wanted authority to fly under instrument flight conditions, I also had to prepare training manuals, train pilots to do the required instruction on the different aircraft types, and convince DOT that we were capable of IFR operations. We also had to acquire aircraft manuals for the different types of aircraft in use and set up a ground school for new pilots as well as refresher training for the existing crews.

These are all normal chief pilot responsibilities. In addition to interviewing new pilot prospects I handled the high volume of paper work required to satisfy the government. This was all very exciting. I was happy in my job and enjoyed learning to bomb forest fires, first under instruction from Al Kydd on the TBM and then from the man who wrote the book on fire bombing, Alexander Linkewich. When we were based together at Kamloops he taught me how to operate the A26 safely and efficiently. Link, now called Link Alexander, wrote two books on fire bombing, the last one, "Air Attack on Forest Fires", is a must for neophytes in the business.

Unfortunately, some of the employees in Conair felt too much was happening too fast and decided they had had enough of my services and enforced regulations. I was abruptly fired after just less than two years in the position. A pilot I had hired to fly a DC6, Peter Mitchell, was my successor. His background was similar to mine and he was a tough enforcer. He was also dismissed sometime later, also without notice. Peter was chief pilot from 1980 to 1983. It seemed that two or three years was the limit of any chief pilot who was not a yes man. As we had enjoyed the work, we both eventually joined another organization called Airspray Ltd. providing the same forest fire service on contract to the Province of Alberta. An unhappy year passed in between, however, during which I licked my wounds and suffered the humiliation of losing a high profile job and working for smaller disorganized carriers in an effort to rebuild the pride and confidence so necessary for a pilot to resume his career.

I am proud to say that no pilots were killed on flight operations during my tenure as chief pilot with Conair. Safety standards, unacceptable to some, were enforced in a way that set a standard that was appreciated by the professionals, but was restrictive to some of the employees who missed the macho image they had been cultivating. Today, flight operations in Conair are very professional and I would like to feel that I was instrumental (and Peter Mitchell after me) in starting the company in that direction.

In 1979, Conair purchased a very successful company from Watson Lake, called Frontier Helicopters, to supplement the fire bombing operations. More and more forest fire control is done today by helicopters. An ill-conceived subsidiary venture into air freight, called Swiftair Cargo, using early DC8-62s, soon went under. The revenue barely paid for the fuel these early fuel-guzzling engines burned. In more recent years, Conair has developed a manufacturing wing converting aircraft to fire bombers. The F-27 fire-bombing conversion was not a success. It was in this aircraft that Ralph Bolton was killed on a training flight in France with *Protection Civile* in 1988. A more recent fire-bomber conversion was named the Firecat. It is a modified Navy Grumman Tracker, used as a company replacement for the Douglas A-26. The installation of turbine engines to the Firecat at great cost was a very questionable success.

Canadair Ltd. of Montreal converted their CL215 water bombers to turbine engines. Although they seem popular overseas and in the Province of Quebec, some still question the advantage of turbine engines on fire bombers. Crews, who have spent many hours behind the noisy, dependable old radial engines, are also reluctant to make the change. There is a great feeling of security in pushing up the throttles and listening to the snarl of the big radials as they power up INSTANTLY. The turbine takes its time spooling up to produce more power. But when you are low and slow on a major fire, you don't always have too much time to waste. In a piston-powered bomber, when you experience a downdraft or have under-estimated the height of the hill ahead, quick power is your lifesaver. Since my time, both Airspray and Conair have been using converted turboprop aircraft with some success. The conversion of a Boeing 737, however, was discontinued. But having been out of the business for several years, new ideas are constantly emerging and I admit a prejudice for round engines.

The Airspray Gang —From left: Mike Hogan, Butch Foster, Al Mac-Nutt, Chris Chandler, Hai Tran, Peter Mitchell, Tony Blake

A Rogues Reunion — Life after Airspray kept the flying friends together. Dining here are Doug McDonell, Dwayne Olson, Al MacNutt and Peter Mitchell.

Bullseye on the Chopper

Where there's smoke there's ...???

One problem Canada shares with our American neighbours and southern Europeans is the hazard of wild forest fires. In Arizona, Texas and California fires can occur any time of the year, but in Canada they are normally restricted to June through early September. Each province mounts its own fire protection system, some more elaborate than others. Some provinces purchase aircraft of their own, while others hire commercial aircraft companies to provide both fixed-wing and helicopter services to combat fires from the air.

In the 1980s when wartime aircraft were becoming scarce and engines to power them no longer built, Canadair Ltd. in Montreal decided to design and build a water bomber. The CL215 was designed to skim water from a lake or river nearest the fire and drop its load to cool the fires until ground crews could get in to extinguish them. The CL215 was meant to replace the venerable Canso, which had served the purpose for many years. This is not meant as a put-down of the Canso - it was old and slow and parts availability poor, but it could outlast many new aircraft. Newfoundland is the only area in Canada where Cansos were still widely used for fire control after their retirement elsewhere.

Fire bombers such as the Grumman Avenger, Douglas A26 Invader and later Douglas DC6s, were used effectively to haul chemical-based retardant. Helicopters are widely used to haul equipment and men to the fire. A helicopter can lower a crew close to the fire by rappelling if no suitable landing pad is available. The men can then quickly hack down a few trees and make a landing area for the chopper when it returns with more men and equipment, including food.

British Columbia has the largest Canadian fire bombing fleet using airline-retired DC6s, converted Grumman Trackers and Convairs. Airspray, in Alberta, had about 17 Douglas A-26s contracted to the Alberta and Yukon governments. The prairies and the Maritimes have less economic need for fire protection. Ontario has a large provincial fleet and Quebec operates the largest fleet of Canadair CL215s in the world. These aircraft do not land at airports except to refuel. They skim along the top of a lake or river to scoop up tons of water in a few seconds and are off back to the fire again. The CL215s are expensive to operate and maintain and cannot be used for any other purpose during the balance of the year. As a result, they are not viable for private enterprise.

In the late 1980s I was flying a Douglas A26 for Airspray. Our crew operated out of Manning, Alberta, a small farm town surrounded by some choice forest that the government wanted to protect. I flew one of four such tankers. Our bird dog officer was a provincial forest ranger in an aircraft flown by Al Hay. As new fires were breaking out in several areas simultaneously, Al was asked to send a couple of us on 'lone wolf' expeditions while he fought the main fire with the other two tankers.

An A26 carries 800 gallons of chemical slurry, which can snuff out a small fire on contact. The chemical eventually erodes with the wind and rain into a harmless fertilizer that promotes forest growth. The slurry itself is tinted bright red so the crews can see where the load drops and correct the next drop accordingly. Normally, we did not drop on the fire but ahead of the burning area.

When Al Hay assigned me to lone wolf a fire reported in a gully north of the Peace River, I reloaded and promptly found the fire as reported. To be named as a lone wolf is an ego trip to a bomber pilot. It implies that we have the knowledge, experience and judgment to find and squelch the fire without benefit of a bird dog and civil service supervision. Thus, when I found my assigned fire, I circled carefully making sure there were no people, cattle or buildings that might be damaged by the drop. A load weighs over four tons and could kill a man or flatten a truck. Confident, I set up my run into the wind and dropped my load on the smoke.

As we had to do our drop at over 120 knots, or about 140 miles per hour, we didn't always get a very good look at the target or see it for long. You are always entitled to a bullseye when you are alone and when the day was over I told my *veni, vidi, vici* story to the group and claimed my bullseye. However, forestry officials were underwhelmed. It seems that Alberta Forestry Service had recently acquired a brand new state-of-the-art helicopter that had just been delivered to them at headquarters in Edmonton. I believe it was a Bell Jet Ranger. The senior brass had decided to take a tour in the new toy before it went into service. They flew to Peace River where they knew they could see lots of fires. For some reason, they had pranged the chopper on the edge of a narrow gully north of Peace River. The aircraft then, unaccompanied by pilot or crew, fell or blew over the lip into the canyon and started to burn. Radios alerted the airborne crews and, unaware of origin of the smoke, our hero arrived and unloaded his bullseye salvo on the remains of a new chopper piled up and smoking in the canyon. Any damage not done already by its unscheduled decent into the canyon was corrected by four tons of slimy red slurry dropped from 50 feet at 140 miles per hour. Bullseye!!

Fire bombing seems to attract a certain type of pilot, some quite long in the tooth. They enjoy the hectic pace of dropping loads of slurry on fires from low altitude in bad visibility and smoke. More important is the ability to be able to sit idle for days between fires. It is probably the type of work that best depicts the adage that flying is "long periods of boredom, intercepted with short intervals of panic". Some skill is acquired over the years in bombing accuracy, which is very much a heads-up operation. The aircraft has to be slowed down to minimum safe flying speed, keeping in mind that a big fire creates its own weather and winds. Sometimes violent updrafts and smoke obscure the target, and we are expected to put our load in the most critical area with only a small margin of error. And, of course, we have to watch the trees in our peripheral vision. Often a dead 'snag', with few if any branches, will stick up as a hazard 30 to 50 feet above the canopy. Being dead they are usually grey and not easily seen in smoke when you get down to 100 feet or less at a speed just above the stall. As we are moving over the ground at about 200 feet per second, we need practice to ensure the accuracy of our drop. An increase in our airspeed, altitude, windspeed, or type of chemical being dropped, will all affect our accuracy and make a bullseye more difficult. Crews are normally allowed a practice their trigger accuracy once a week when there are no fires.

The quickest way to get into trouble with a heavy aircraft at 100 feet or less over the forest at minimum speed, is to take a quick look in the cockpit to check something, or to answer a radio call and lose your concentration. Radio silence is important when a bomber pilot is on a drop. Idle chatter on the radio is not welcomed.

Normally, the drops in Alberta are on relatively flat terrain and the pilot can often see the target smoke from many miles back. When the line or direction of drop has been established by the bird dog he circles the fire for what is called a lead-in. He will have made a series of dummy runs over the fire to ensure there are no snags on the line and to establish that the bomber will not be running uphill, making the overshoot dangerous. Any heavy smoke and turbulence will be checked to ensure that they are not major hazards. The bomber will approach about 800 feet above the target in a left hand orbit for best visibility and then descend and release either one or two doors in a line - a string - or a salvo of the whole load at once, as directed by the forestry officer in the bird dog. A full load is about 800 gallons (8,000 lbs.)

In flat land Alberta this is not a great problem. Accuracy improves with practice. However, in the foothills of the Rockies, where many fires start on both sides of the Continental Divide, lightning fires can cause great damage. Much effort is spent to control these fires in both Alberta and British Columbia. In the mountains, the tree level is usually up to about 7,000 feet. Above that there is nothing to burn, although the mountains go up to 12,000 feet making access to the area difficult. But when the fire is on a steep hillside, access by tankers requires special skills and intestinal fortitude.

ALTIMETER RISING

Often the fires are hard to find for the lumbering tankers. You can be flying up a valley only a few miles from the fire, which is often the other side of a ridge. Not only can you not see the fire; you cannot speak to your bird dog for instructions, as your VHF radio communications are only line of sight. The messages do not go through the granite hills any easier than does the aircraft. When contact is established, the tanker pilot must climb to the top of the ridge, and stagger along at low speed, at a minimum safe height over the rocks. Then, when he gets to the proper canyon, he does a sharp turn and dives down following the terrain, often seeing his target for the first time and only briefly. His speed increases rapidly due to aiming 35,000 lbs. down a steep hillside. He must drop his load early, as it will do little or no good behind the fire. Fires go uphill very quickly. The good part of bombing in the mountains is that you have lots of speed at the bottom of your run to dive out and avoid any obstacles in your departure route. The downside is, if you don't get clear of your load you can wind up at the bottom of the valley with a heavy aircraft and a lot of climbing and turning in a tight area is required to get back up to the ridge for another run.

Veterans like Al Kydd, Tom Wilson, Peter Mitchell, Rod Boles, Brian Johnston, and Link Alexander, who have done this for years, find it a snap, but it is no business for amateurs. 'Rotten' Ralph Bolton, from Mission, B.C., no longer with us, was one of the best fire bomber pilots I have ever watched in action. Other pilots who might think fire bombing is a glamorous occupation, a concept supported by the movies, look for a career change after a summer sitting in the rain with no fires.

Captain Bob Learns the Trade

DC3 to Airbus – the next generation

Few things give a professional pilot more pleasure than initiating others into his craft, and when the trainee is your own family the enjoyment is doubled. When our kids were born, my wife and I had a temporarily successful flying school in La Belle Province and our kids spent most of their pre-school years on the grass infield watching Tiger Moths, Fleet Finches, Aeroncas and old model Stinson and Taylorcraft doing 'circuits and bumps.' As they grew older, father's flying was farther afield and for many years I would only spend three or four months at home while flying locally. The rest of the time I was based somewhere in Africa, Spain, Chile, Venezuela, or wherever our company's contracts required. A lonely period for survey pilots and for survey wives.

Captain Bob MacNutt in the cockpit of his Canadian Airlines 737.

As my seniority and responsibilities increased, I got more time at home doing pilot selection, training and ferrying the survey aircraft with a fresh crew to some foreign project. I would get the crew settled in and started to work, bribe the local functionaires into a co-operative stance, and return home to my family.

My wife had learned to fly in 1947, and our children, surrounded by pilots as our peer group, were more at home around airplanes than around cars. Even my nephew, George, lured by the airplane stories, came to live with us and left with a commercial pilot's licence. Now retired as a 747 captain, he has flown many types of airplanes for many years with the flying goose design of CPAir.. He also has spread the word of aviation by starting several flying clubs, and was a national director of the Canadian Owners and Pilots Association, a national organization of amateur pilots. He flies small airplanes as a hobby.

A cousin, Bob Beer, from Charlottetown, came and lived with us, earned a pilot's licence and an engineer's licence and started a flying service in his hometown. Jean Marie Pitre, a student, became a Trans-Canada Airlines captain, Ben Rivard became Operations Manager of Eastern Provincial Airways, and Leo Lejeune, Chief Pilot for Quebec's fire fighting CL215 program. These achievements have given me great pleasure. But the greatest pleasure of all is to see your own sons learn about the joys and scares of becoming a pilot.

Bob (left) is congratulated by Ottawa Flying Club instructor Bernie Duperron on completion of his commercial pilot's licence.

When the kids were still young, we bought an old Taylorcraft, renewed its fabric in our garage in Ottawa, and taught the three boys to fly. Later, with a good friend Ray Taylor, a CPAir navigator before Inertial Navigation Systems made navigators redundant, we bought a four-place Piper Tri-Pacer, recovered it and moved the kids into the element of instrument and night flying.

When Bob was training for his multi-engine rating and IFR endorsement in the mid 60s, we rented a twin engine Piper Comanche from a local entrepreneur as a training aircraft. We had much fun at this and scared each other several times in the process. On one occasion, I was teaching him minimum control speeds in flight -- the lowest speed possible with one engine out and the other at full power. Below this speed the aircraft will start to roll over on its back - a nasty manoeuvre, especially if you happen to be at low level or in cloud. As his instructor, I should have known that the stalling speed (V_S) and the minimum control speed (V_{MC}) on that particular airplane were almost exactly the same; but no one had told me and we got into a potentially dangerous situation.

As we practised V_{MC} (about 80 knots if I remember correctly), the aircraft suddenly stalled and went into a spin. Five thousand (yes <u>thousand</u> not hundred) feet lower, we got the thing flying again and went home two very shaken pilots. As a result of accidents around the world, the flight specifications on the Twin Comanche were later changed to reflect this problem. Fortunately, we were doing our work at 7,000 ft. above ground level and recovered about 1,500 ft. above the ground, but we were badly shaken and knew something was basically wrong. Had we started the manoeuvre at say 5,000 ft, as would have been perfectly normal, we would have been unable to return the man's airplane.

On another occasion, I was in the left seat and Bob was riding as co-pilot for some reason. When the trainee has achieved a degree of proficiency, you tell him to be prepared for simulated engine failure at any time. The instructor, whether on a single or twin engine trainer, will throttle back an engine to simulate an engine failure and the trainee should be able to handle the emergency as he has been trained. If one engine fails during take-off, assuming there is sufficient airspeed and runway, a pilot should be able to get the airplane into the air, fly a circuit and land.

ALTIMETER RISING

With me driving, Bob decided to do what I had been doing to him, 'pull a throttle' at V_1 and see how I made out. Fair ball, it was time to put my money where my mouth was. We got the beast into the air, but not before a little excursion through the cabbage patch on the side of the runway. We missed a runway light, got back on the cement and into the air. Normally, in such a situation, the action would be to shut down the other engine and brake to a halt on the runway remaining, but once again a pilot's ego got in the way and I had to do it the hard way. At least I proved to my astonished son that 'Yes, Virginia (and Bob) there is a Santa Claus'.

In the early days of aviation, bush pilots in the north were often asked to take out heavy loads from a small lake or cleared strip, often with trees or rocks at the end. These judgment calls by the pilot were very critical. Getting the maximum length to accelerate, the temperature, elevation of the take-off area, condition of the strip are all-important factors. Early pilots tell of tying the tail end of their floatplane to a tree on shore and setting fire to the rope. This gave them time to get in and power up to maximum before the rope burned through. When it broke you were on your way. Running up under full power with the brakes on can be hazardous, with stones ruining the prop or dirt ingested by the engines. But, of course, the luxury of brakes is not one the bush pilot on floats or skis can enjoy.

Sometimes on a short trip we would start gaining speed going downwind towards the take-off button then, at a safe speed for the undercarriage side load, turn gently into wind and add full power. That would give us the edge of a few knots to start the roll. When jets came into production with high gross weights, runways were lengthened and a more scientific approach was necessary to calculate safe take-off/climb/abort and landing approach speeds. The V for velocity was introduced. The co-pilot now calls the V speeds to the captain while he's busy keeping the aircraft on the centreline. This procedure ensures that suitable speeds are being achieved for a safe take-off. Earlier pilots often established a certain point on the runway where they must be airborne or they would abort. They would return to the ramp, unload some cargo and try again. The marker could be a tree, rock or shack along the runway. Later, when turbo props became popular, the co-pilot would count out the seconds on the roll and if the indicated airspeed did not come up to a pre-determined safe level after so many seconds, the take-off would be aborted. Runway condition, temperature, air density, wind factor, and aircraft performance are all important as is, of course, runway length and obstacles on the departure path.

Engine life in the early days of turbo props was based on cycles. It was not important how many hours the engine had run, but how many times it started and stopped (heated and cooled) that affected its life. Nowadays, passengers on commercial aircraft are not always aware that before the captain starts his take-off roll, he knows his gross weight, centre of gravity position, fuel load, weather en-route, destination, alternate and much more. Also he has instant communication with his company as well as ground and air traffic control. He normally has a fax to get updated weather and other en route information. You may be assured that he has been trained on the type, has had sufficient time under observation by his supervisor and his medical and IFR proficiency check are current and valid. He will have discussed with the other pilot one or more simulated emergency situations, departure procedure, headings, altitude restrictions, and tested his oxygen mask flow.

The various speeds they discuss are V_1 the speed at which the aircraft can safely abort within the runway confines. V_2, the speed at which the aircraft can safely climb, and V_R, the speed at which the pilot in command will 'rotate' or raise the nose for take-off. These speeds vary with conditions for that particular take-off and vary with prevailing conditions. Similarly, the arrival briefing includes approach speed at the aircraft's new weight, and a discussion of their go-around procedure if the runway is not sighted in bad weather or the runway is not clear. These speeds are placed on the 'bug', a small cursor on the airspeed indicator for easy reference. Some of the few constants in flight procedures are the maximum speed at which landing gear and flaps can be deployed.

ALTIMETER RISING

One final comment on your jet trips to Hong Kong or Australia. Aircraft nowadays on a long flight burn off thousands of pounds of fuel en route. In some cases, landing weight could be half the take-off weight. Aviation has changed during my 50 years at the 'pole' and is changing and will change much more. Solid fuel, for example, would be helpful to avoid fire in an accident.

When Bob writes his next generation memoirs, he will have much to add. For instance he will not be getting 'pole time'. Modern aircraft have no control column, or pole – just a toggle switch like Nintendo. These days "Captain Bob" practices his exercises from the left seat of a multi-million dollar Airbus simulator, courtesy of Canadian Airlines International, with more precision than he could ever learn from me. In addition to being safer, this kind of drill is much more economical. For these reasons, airlines can schedule training more often, in a more professional setting, with no schedule interruptions and no wear and tear on the airliners. (At time of writing, Canadian Airlines has been purchased by Air Canada.)

In earlier days, several pilots would be herded into an aircraft after it came off 'sked', hurried through the bare necessary training, all after a hard days work and pressed to get the ship back to Maintenance before midnight, or back on the line for another departure. Simulators are beautiful.

After starting his career as a commercial pilot in survey, Bob got a job as a co-pilot on DC3s, based in Northern Quebec. His Chief Pilot was Jean Marie Pitre, a handsome red haired lad I had taught to fly many years before. The timing was beneficial, as when the 'call' came from Canadian Pacific Airlines, every young pilot's dream in those days, he was asked to do an evaluation flight from the right seat of a DC3. He had paid his dues in that seat and was ready for the career move. That was 30 years ago, and now our Captain Bob pays more income tax than I ever made as a pilot! And he enjoys flying and is teaching his kids.

The bush pilot part of his career served him well, as he is keenly aware of how cold it is outside the airline environment, especially during our 'recessions' when the corporate jet is the first to go. When you have flown a winter on wheel-skis in the sub-Arctic, rolling gas barrels in the snow, freezing in tents at night, and eating cold meals from the can, you can appreciate the four stripes of an airline captain better and wear them proudly. The Canada goose symbol under which

Jack MacNutt (left) flew with Sheldon Luck during fire bombing operations out of Thunder Bay. Jack is now an air traffic controller.

he has flown for many years has been stylized, along with the procedures of flight. But he still enjoys flying.

It would be difficult for a non-pilot to know the pleasure I got recently when I had a chance to watch Captain Bob at work. My wife and I were en route to Halifax on vacation, and were fortunate enough to get on a 737 captained by our son. When invited into the cockpit to watch him operate that beautiful jet in a very professional manner, the joys of parenthood were very profound.

Son Jack also holds a commercial pilot's licence, but failed the eye test for airline standards and, after several years as a pilot in general aviation, he trained as an Air Traffic Controller. Jack hauled fish in Cansos, a smelly job on a hot summer day; he flew as birddog pilot for aerial fire control, and later spent two years flying an A26 fire bomber. He is now studying for his instructor's licence and hopefully will pass on his skills to others for their enjoyment in the air. In the meantime, he is an Air Traffic Controller and private flying enthusiast in Prince George, British Columbia.

Son Jim, also an Air Traffic Controller, is an ardent pilot, but prefers to fly for enjoyment. His kids are also keen aviators, so Pappy feels he has done something right. Jim's ability to put words together in a logical and interesting fashion far exceeds that of his father, and he has written many trade magazine articles over the years. He follows current developments in aviation and space technology very closely and hopefully will one day write the sequel to this humble book bringing aviation readers into the twenty-first century with all the amazing gadgets and airplanes that will be forthcoming. He was hauled to the airport, and in and out of airplanes, before he was toilet trained and thus already has 40 years of aviation lore.

Those of us in aviation who have watched it grow and produce men like Max Ward, Grant McConachie, Carl Burke, Wop May and Weldy Phipps, wonder if we shouldn't recycle our politicians and keep the aviation industry intact.

They fly better than they golf, Al admits. From left: Bob, Father, Jack and Jim.

ALAN MACNUTT

My Love Affair with the Canso

Life is too short to fly ugly airplanes

The aircraft that evokes the most memories, and which is probably the greatest cause of my present hearing difficulties, is the amphibious Canso. The Canso was designed by Consolidated Vultee. Most Canadian Cansos were built by the Boeing Aircraft Co. in Vancouver during the Second World War. It is a high-wing monoplane powered by two 18-cylinder, twin-row Pratt and Whitney R1830-92 piston radial engines driving Hamilton standard 3-blade, full-feathering hydromatic propellers. The power, the propellers and the carrying capacity are similar to a Douglas DC3, but there the similarities end. The Canso's propeller tips pass very close to the pilot's head, the engines are close to the fuselage and soundproofing

Docking a Canso was a tricky operation, as in this B.C. maneuver.

is not something anyone worried too much about in wartime. The noise level in the cockpit is horrendous. and it is very little better in the cabin.

The wing and engines are located high above the landing surfaces, as in all amphibians, to prevent water damage. Huge ship-style girders are built into the cabin to give the structural integrity necessary for landing and taking off in rough water. Since the Canso was designed in the 1930s by a naval architect, it has flown in many roles in many countries, and is still in active use in many parts of the world. A few of these applications follow and I am sure there are many I have missed:

- Submarine patrol
- Water bomber (forest fires)
- Rich man's yacht
- Executive transport
- Ice patrol
- Fishermen's and fish transport
- Hauling wild rice
- Offshore drill crew transport
- Photo survey
- Magnetometer and electro-mag survey
- Engine test bed
- Glider towing
- Military transport
- Movie camera platform (Jacques Cousteau)
- Search and rescue
- Hinterland exploration

In addition to the high noise level, uncomfortable cockpit, and dawn-to-dusk survey flights, the Canso had a third strike for survey crews -- it was a pig to fly. The controls were very heavy and required a high level of physical strength. As the autopilot was rarely serviceable and unsuitable for survey flying in most cases, the co-pilot had lots of opportunity for hands-on flying -- often more than he had bargained for. Although all the above tends to downgrade the Canso, and initially most pilots' reaction to the airplane would be unprintable, the character of the beast grew on you. After a few hundred hours, you grew to actually enjoy the airplane. It had a reliability and sense of structural integrity that you grew to trust. After you had mastered a Canso, and docked it a few times and landed and taken off in water conditions that would swamp a lesser flying boat, you felt that anything less was child's play. A perfect image for the macho ego conscious pilots of my day.

Younger pilots, spoiled by modern jets and turbo props, might point out that the old Canso has no flaps, no secondary hydraulic system to take over when a pump fails or a line ruptures, the nose wheel does not always come down, it is a fuel hog, parts are hard to find, and it is ugly, slow, cold, and vulnerable to icing. Lacking experience on type, they might overlook the noise. Unfortunately, all this is true, but we Canso addicts insist it will do specialized jobs of which no other aircraft in the world is capable.

Originally designed as a PBY, the 'Catalina' had no wheels and was strictly a flying boat. And boat is the operative word. It was designed to withstand rough usage and has proved very forgiving in many ways. It has one characteristic that seems to stem from its original design as a boat; it will not tolerate being landed on water with the landing gear extended. To my knowledge no crew has survived a gear-down landing on water. The impact shears the 'tower' which joins the large overhead wing to the boat part, crashing it down through the cockpit. However, it will withstand landing and taking off in stormy water conditions that most large boats would find unpleasant. That is a statement I can vouch for personally. Some 3,000 Catalinas were built, but only 615 were wheel-equipped as Cansos.

One of my jobs as a Department of Transport employee was to conduct airborne rides on pilots for their instrument rating renewals and proficiency rides on the aircraft type (PPC/IFR). This constituted a preflight briefing to demonstrate a knowledge of the aircraft, a flight test covering some basic air work and two or more instrument approach procedures in real or simulated bad weather conditions.

Rough water was not the only hazardous landing surface for the robust Canso. Ray Bernard proved that in another dimension. A veteran pilot of Dutch origin, Ray owned a Canso and a DC3 at Edmonton. The Canso was not used much in winter as it was really cold in the cabin and there are few fishing parties wishing to be flown into the north from Edmonton at that time of year. I was scheduled to do a DOT check ride one winter afternoon with Ray in the Canso. Ray checked the oil, added some fuel, his engineer 'Frenchy' Lamoureux got into the 'tower' (as it was a three man crew) and, with a co-pilot in the right seat, I observed the flight progress while trying to stay warm.

Bernard was an excellent pilot with lots of experience on the Canso, but I still had to witness at least two acceptable, instrument approaches before I could give him a pass. As he had not flown for a while, he made a couple of errors and we had to redo an ADF approach at Edmonton International to meet the requirements. The aircraft was based at Edmonton Municipal, but it was advisable to use International for testing due to better instrument facilities and less traffic.

An ADF approach is an instrument approach, without visual reference to the ground. By using one or more non-directional beacons on or near the runway the pilot can find his way down through the overcast to visual contact with the runway in time to conduct a safe landing. Unlike an ILS (Instrument Landing System) approach, which is much more precise, the ADF approach is simpler and much less accurate. It is, of course, cheaper to install and maintain in both the ground and aircraft installation. ADF beacons have been in general use at smaller airports and in the north for many years. They are still available even at major airports, but used only as a backup.

As Ray did his procedure turn for what we hoped was the last approach, one engine died for no apparent reason. As there are no meaningful fuel gauges or fuel selectors on the Canso we opted to land right away and investigate. As he rolled out on final approach about two miles back, the other engine quit and we had a very quiet high wing glider with no flaps or spoilers but a very large wingspan.

As my services were not required, I went aft and buckled myself in to await the effects of gravitational pull. Ray made a beautiful landing 'dead stick' in a farmer's field which had a good crust of snow hardened from the winter long Alberta winds. Gradually the left wing dipped as the aircraft slid to a rest and the Canso sat proudly on the snow drifts resting gently on the left tip float. The wing tip floats had been lowered but, of course, the gear was still up as we had no hope of reaching the runway.

It was dark by now so we boarded a snowmobile driven by the farmer who owned the field and made our way home for the night. My duties took me elsewhere in the morning and I did not see the take off, but Ray sent me pictures which showed the old Canso, with fuel added, lifting off the snowbanks, rising gently over the fence and home to Edmonton Municipal absolutely unscathed. When I saw it a few days later I could barely make out a spot on the keel where the paint has been removed in the snowdrift. Bernard's explanation for the double engine failure was a wiper rag that got into the fuel system and blocked the fuel flow to the engines. He never did forgive me though when I failed his ride and we had to do it again -- with fuel.

In later years, Ray chose a warmer climate with no snow. He spent his winters flying rich tourists up and down the Nile from a base at Harare, Zimbabwe. Passengers could visit Zambia, Tanza-

Ray's embarrassed Canso takes off without wheels from the snow.

nia, Sudan and Egypt to explore famous game parks, historic sites and points of interest. The aircraft - the ubiquitous Canso, of course.

During this same period with DOT, I was assigned to do a route check on Austin Airways' route up the east coast of Hudson Bay. The Canso captain was Peter Sosnowski, who later worked for Northwest Territorial Airways out of Yellowknife on their DC3s. Our route was from Timmins to Moosonee, along the south west shore of James Bay to Fort George, Great Whale River, Port Harrison and to Povungnituk on the east shore of Hudson Bay and return next day. It was a trip of about 2,000 miles in an airplane cruising at 125 mph. The cargo would all be hand carried and delays were to be expected everywhere. This was, however, an important route for the people in these communities. This was their once a week contact with the outside world; their mail route, food supplies, outboard motor parts, etc.

Out of Timmins I did my normal thing of moving around the Canso checking such items as emergency equipment, exits, and cargo security. As our flight gradually worked its way north I got into the paperwork and found the aircraft did not have a valid Certificate of Airworthiness. The C of A is good for one year but had lapsed two months previously. The Certificate - now outdated - stated that a major inspection had been done and it had been certified as meeting all the safety standards by a licensed engineer. This is considered a very basic no-no for the inspectors, especially on a scheduled passenger service by a reputable airline. One of the first considerations was that the flight was illegal. What to do?

Farther south it would be a simple decision to ground the aircraft at the next stop until it was legal to fly again. In Great Whale River, (later called Poste-de-la-Grande-Baleine, now Kuujjarapik) these passengers, mainly Eskimos, many with their babies and families, would be stranded. It was not a matter of walking across the terminal and continuing their flight to destination with another carrier. There was no other carrier, or terminal. This aircraft represented shelter, food, and a home for the natives until they got back to their destinations in the far north.

When we landed at Great Whale and Peter got his crew busy unloading. I took him for a walk and explained he had no C of A. Like myself he was in a tough position. He risked a violation for flying an unserviceable aircraft and, far worse, in the event of an accident. This was not the usual confrontation of brown briefcase bureaucracy versus commercial carrier considerations (known in the trade as BBB versus CCC). It was more a matter of consideration of what would happen if the innocent passengers were left on the beach at Great Whale while someone shuffled paper. The aircraft had been signed airworthy that morning for its daily inspection by a company mechanic. It was obviously in good condition and the crew very professional. We agreed that Peter would contact Austin and have them request a temporary C of A and have the dispatcher authorize his continuation of the flight. Meanwhile, I would do a CYA manoeuvre by formally advising the carrier of the problem -- after the aircraft had departed -- and I would stay behind and not appear to actively condone the breach of air regulations. The outcome spoke well of Jack Austin's prestige and the respect given him by the Department of Transport. That evening there was a wire sent ahead to the crew, with a carbon copy to me as the air carrier inspector responsible, confirming a new C of A had been issued that evening. The amazing thing was that the wire was dated 6 p.m. and it was Friday! I had been trained that the world stopped in DOT at 3 p.m., if not earlier, on ALL FRIDAYS!

In 1980, after a two-year stint as chief pilot for a fire bombing company on the West coast, and a brief period as consultant for some smaller companies in the Vancouver area, I once again went back to sunny Alberta as chief pilot for Avalon Aviation out of Red Deer, and was back on the Cansos, this time as a water bomber pilot. Unlike a chemical carrying aerial firefighter, which must return to the airport and reload after each drop, the water bomber, can skim over the surface of a nearby lake, scoop water up in a few seconds, and return to the fire for another drop. An internal tank adds a foam mixture to the water to give it adhesion. This system works well to cool a wildfire and hold it until the ground crew can control it, or put it out. The water evaporates quickly, sometimes before it even hits the ground. Water bombers are considered cheaper to operate, can deliver a vastly larger volume per hour due to the short turnaround time. Sometimes, when there is a suitable lake adjacent to a forest fire, you can reload and return in 10 minutes or less, and it is not unusual for a water bomber to deliver 60 or even up to 100 loads a day for a big fire.

The Canso was the principal bomber in the 1960s and '70s, but was later replaced by more efficient land based tankers and by the Canadair CL215, the only aircraft in the world designed for the sole purpose of fighting forest fires.

I have always liked to think of myself as a teacher and instructor. I was therefor pleased that one of my responsibilities was to hire and train Canso crews. That year, we had contracts in the Northwest Territories and in Ontario during their fire season. I had one mid-season challenge that stands out in my memory. One of our most experienced bomber pilots, Bill Cassleman, got a job on Canadair's new CL215. I had 72 hours to get a replacement to Thunder Bay. To replace a first officer on a specialized operation is easier. He only has to be qualified on the aircraft and the captain can teach him the skills required.

The only experienced Canso Captain with a background in fire bombing I could think of was Sheldon Luck, former chief pilot of Flying Firemen, a competitive company. Sheldon had retired twice previously and was busy writing his memoirs in northern B.C. My boss concurred and the two of us spent an evening on the phone, taking turns talking Sheldon into going back to flying at age sixty-six.

My son Jack was a first officer with Avalon at that time and he often recalls the experiences he had flying with Captain Luck, who is a legend in Canadian Aviation. In the early days, before it became Canadian Pacific Airlines, Sheldon was Chief Pilot of the company. Ron Keith has written much about this era in his biography of Grant McConachie, *Bush Pilot with a Briefcase*. Unlike more insecure pilots, who are reluctant to share their trade secrets, Shel was always willing and ready to help a younger man learn the trade, a characteristic that I admire in pilots and a habit I like to think I share.

My boss during this period was Bob (Moose) Murdock, a veteran pilot. Moose, unlike many pilots, did not have any urge to write a book. In fact, he had probably never read one, but he was a colourful character and his aviation stories would make great reading. Kirk Carleton and Darryl Murakami were two of our engineers in those days and I had the pleasure of working with them both again 12 years later in Airspray. Kirk was our engineer and our maintenance was excellent. He had a gentle, but persuasive way with his crews and, as a result, he brought everyone up to his own high standards. Not being a pilot, he did not suffer the ego trips we pilots endure, and which eventually result in our humiliation.

As my entrepreneurial tendencies kept surfacing, I returned to the West Coast the following year and joined a group of businessmen in acquiring a Canso. We obtained authority to operate a charter service up and down the 'Super Natural' B.C. coast carrying sports fishermen from Vancouver International Airport. We booked parties of up to 24, mostly Americans and Germans, and flew them into the fishing camps and to charter boats serving the coastal area.

Some of our weekly trips were to a fish camp in the Rivers Inlet fjord. Others went to Hakai Pass, where we would find the 'Huntress', or to Shotholt to take a load into Claude Lacerte's ship, the 'Belle Blonde'. The 'Thorfin', cruising near Calvert Island, was also a customer and the Good Hope Cannery Lodge was a regular call in Rivers Inlet. The cruise vessels catered to parties of 24 and provided first class accommodation while the patrons fished, drank, played cards, fought, and escaped their wives and responsibilities for one whole week. At week's end, we would bring in a new load with their fresh supply of booze and bring out last week's group with their fish, tall stories, and incredible B.O.

Unfortunately, we rarely got time to fish ourselves, as we did two return trips daily, weather permitting, all summer. Once a week, always at night, we would deadhead into Victoria after the last trip and hand the aircraft over to Russ Popel of Victoria Aircraft Maintenance. His crews would work all night so that we had the aircraft ready for service at daylight the next morning. I was the only one of the owners who flew, so I was in charge of flight operations. I was assisted by Capt. Bill Hinds, Gisella Barry, chief stewardess, and Linda Gawda. We had a very busy summer. Our dispatcher was Bill Bevan, formerly of Trans-Canada Airlines, and young Angus MacDonald, a co-pilot, son of a judge in Langley, B. C. Bob Dyck helped out when we needed a hand. A former RCMP pilot joined us briefly, but not being used to flying in bush pilot weather, and operating older aircraft at full gross weight off rough water he did not stay long.

The Canso was awkward to manoeuvre on the water as both engines were close to the centreline and, of course, there was no water rudder. It was particularly difficult to dock at a wharf and we would anchor near the fishing camp or yacht we were servicing and transfer the passengers by small boat. After a couple of hours en route in the Canso with our stewardess serving free drinks, many of our passengers had to be lifted into and out of the boats. Lifting 24 inebriated, overweight sportsmen into a small boat from a Canso hatch and helping 24 back in, was the hardest part of the job. It was harder still to smile about it and praise their catch of salmon.

Bill Hinds, a handsome young lad of Scandinavian descent, was the other captain and we enjoyed working together. One day Bill and I were flying together and the weather, always unpredictable in Hecate Strait, closed in suddenly. We could not seem to get out of the clag. The mountains at the mouth of the fjords are high and the Canso could not just climb up on top like a conventional airplane. Cansos are boats and have never been trained to climb. As our visibility deteriorated, and some small islands and ships were much too close for comfort, I decided to land straight ahead in the Pacific and wait for better visibility. The surface from 200 feet looked calm enough. When we were about to touch down, a wall of green ocean leapt at us and bounced us off with an unhealthy resonance. The next swell was even worse, but by the third one the Canso had gotten discouraged trying to fly and flopped into the trough like a wounded goose.

We checked around and the parts appeared to all be in place. I went back to check the cabin to see if we were about to sink, while Bill held the aircraft into the storm with the engines. Gisella Barry needed all her charm to convince the passengers that all was well and we would be airborne as soon as the visibility improved. I was glad to hear it myself. After an inspection for leaks and lacerations, I was glad to get back up front. Gisella doubled the drinks to keep the passengers from asking too many embarrassing questions. The fog lifted briefly and, as we were bobbing about like a cork, we had the passengers strapped in and took off.

The take-off into the swells was more exciting than the landing because it was noisier and we hit the mountains of water at higher speed and with impacts that would shatter most aircraft. Back in Port Hardy a couple of hours later we could not find one rivet or a panel loose. The only damage was to the pilot's ego.

Experiences like this gave me a great appreciation of the many bush pilots who had been doing this sort of thing for years using the

The Canso crew that hauled fishing tourists up the west coast of British Columbia included Linda Gawda and Gisela Barry, (kneeling) Bill Hinds, George Filiatrault and Al MacNutt.

Grumman Goose and seaplanes long before we had modern airports everywhere. People like Stu Spurr, Al Eden, Johnny Boak, Wally Russell, Jack Ross and many others, made a career of operating in this beautiful environment where the weather is fickle and passengers are demanding. Justin de Goutrie wrote a book on flying in this area, called 'The Pathless Way', which is regrettably now out of print.

I have flown commercially in many countries on five continents and never have I seen weather more unpredictable than the area between Vancouver Island and Alaska. Scandinavia in the winter time has much fog, but it is predictable and consistent. Africa has the sand storms that ruin visibility and engines, but last only a few days. The West Coast is unpredictable at any time and has an angry ocean on one side with steep mountains on the other, and occasionally some very nervous pilots in between!

The sad part of the story is that the president, one of the partners, left town suddenly at the end of the season with our summer's receipts, which were quite substantial, and left us with unpaid fuel and maintenance bills. We owed several payments on the Canso, which was promptly seized. It has been said many times that pilots are poor businessmen. The humiliation is not being able to pay people what you owe them when they have worked hard to help you succeed. Incidentally, the president who ran with our cash was an American and this was my introduction to free trade long before Mulroney made it all legal.

One young man who may be on the right track, is Dave Dorosh of Edmonton. Dave, who has spent most of his career working as a supervisory maintenance engineer on Twin Otters in Colombia, bought a war surplus Canso about 1960. Instead of putting it in service or leasing it out as he was urged to do, he put it in a hangar at Gananoque, Ontario, where it still sits forty years later. His fixed costs are low, operating costs are nil, and the aircraft becomes more valuable daily. I'm not the only guy who has a thing about Cansos!

The long straight wing of the Canso produces negative yaw. For a non-pilot, that gobbledegook means that if you start a turn to the left in a Canso by applying aileron, as you would in a normal airplane, the nose swings to the right. In the Canso you feed in a lot of left rudder and, sooner or later, it starts to go the way you want -- to the left. In rough air, when you are being tossed about, the Canso can be very tiring to keep straight. It requires strong physical co-ordination inputs for a smooth ride. The venerable DC3 shares this switchy-tail business, and has given many passengers all over the world the tendency to join the lineup for the can at the rear. And the last thing the pilots need is a tail heavy DC3. Robert N. Buck, a retired 747 Captain with over 50 years flying experience, explains the 'Dutch Roll' so noticeable in the Boeing 707 in his excellent book 'The Art of Flying'. Yaw dampers on modern airliners have resolved this problem, common to earlier aircraft. This not only decreases the work load of the crew, it makes the ride for the rear passengers much more comfortable.

One more item you should know before you are fully checked out on flying boats, is most notable in the Canso. In a conventional aircraft, if you should reduce power in cruise the nose will drop. Conversely, if you add power, the nose will tend to rise. In the Canso, as in most boats, the weight is <u>below</u> the thrust line. When you reduce power, the stability, or more correctly the equilibrium, of the aircraft is upset. The bottom tends to move ahead before the momentum bleeds off and the decrease in power <u>above</u> slows down, immediately pitching the nose up in the air. An increase in power in cruise has the opposite effect, pushing the nose down until a new equilibrium is stabilized. So, to confuse everyone, if you want a Canso to yaw to the right, try and turn left; and if you want the nose up, reduce power! I told you the Canso is different!!

Boarding passengers at sea was not an easy task, especially after they had a hard day's fishing.

ALTIMETER RISING

A Tribute

Upon Al MacNutt's retirement from commercial aviation, his three sons presented him with this montage of the many aircraft he flew in his 50-year career.

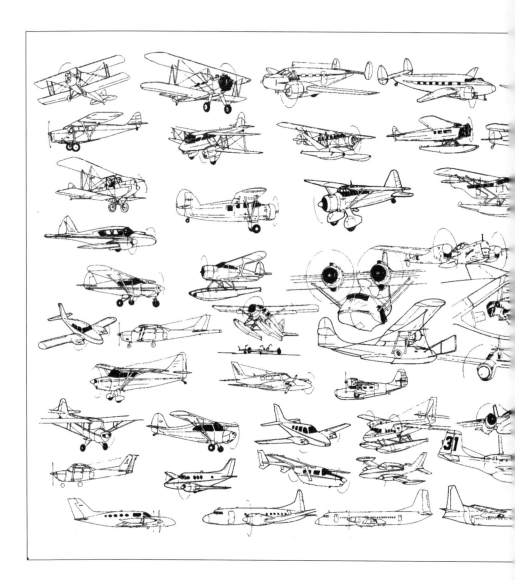

ALAN MACNUTT

50 YEARS

W.A.(AL) MacNutt

He found his element,
In the air.
And endowed us with it.

He understands –
 –the machines of flight,
With lifted intimacy,
And shared it with us.

He showed us, McGee's,
 "Footless Halls".....
And we danced on his –
 "Laughter – Silvered Wings"
He set us free.

We honour him;
As our father.
As the true –
 Consumate flyer.
A pioneer, rebel, visionary
A pilot's pilot .,
Thanks Dad –

Bob
Jim
Jack

Time to Move on

The completion of an ideal career

I have had the pleasure of flying, almost without a break, since I joined the Royal Canadian Air Force in 1942. As I look back on my 50 years in the cockpit, I count myself lucky. Few people, least of all pilots, are being paid to do what they enjoy doing most at age 68. On this anniversary date in July 1992, one of my respected colleagues also hung up his spurs after a half century in the cockpit. Harry Chekaluck, a fellow Airspray pilot, flew Lancasters in Europe during the war, worked with smaller airlines later, and then worked for the last twenty years of his career with the same company. Good luck Harry! It's too bad we can't pass along all we've learned between us in 100 years of aviation to these young bucks. But, of course, they wouldn't listen. In any case, they need to make their own mistakes. It's the only way we seem to learn and remember. Let's hope their mistakes, like ours, are not big ones, and let's hope they get the same enjoyment out of aviation as we did. Let's hope as well that they are privileged to fly as long.

I am amused to remember that I often wondered how my camera operator -- an 'old man' of fifty four -- could stand the physical exhaustion of working in an unpressurized cabin at about 38,000 feet. The arrogance of youth! At sixty-eight I was still flying a comparable type of aircraft albeit at a much lower altitude. I guess it just depends on where you sit. I'm sure some of the young bucks around me then wondered if I was due to be carried out next week because of my old age. I, in turn, looked at my training captain, a

Airspray boss Don Hamilton wishes Al well on retirement day. It was tough to leave a job he loved for 50 years.

young lad who just learned to shave, and wondered if he was experienced enough for this job.

It's time for me to move along too. Several young men are hinting that they would like my seat on beautiful old No. 31, my Douglas A26, and the check pilot gets more unreasonable each year. For some strange reason, he thinks he can fly the aircraft better than I can, which I find hard to understand. But then, every pilot thinks he is the best.

Before I disappear off into the sunset to fly at my local flying club, let me tell you a little about the aspect of aviation Harry and I have been involved in for the last years of our careers. Our boss has been Don Hamilton. He sold the old Anson in which he had hauled fish for a living and bought Airspray in 1967 for pocket money. He took a contract to operate a Douglas A26 fighting wild forest fires in Alberta. Twenty-five years later, Don operated eighteen A26s, about ten light twins for bird dog aircraft, and a Cessna Citation business jet that he flies for pleasure with his daughter, Janice. He owns them all. Don and Janice also operate a fleet of Canadair CL215s for the government. It is no secret that Harry and I have been permitted to fly into our golden years because our boss likes to fly as much as we do. And since he is the same age, he doesn't see chronological age as a big problem.

I've already told you about the A26 and what a gentle aircraft it is to fly and how it is simple to maintain and operate. I've explained the lovely racket the big radial Pratt and Whitney 2800 engines make and how they snarl up to 2000 horsepower each for take-off. They haul eight thousand pounds of fire suppressant chemical at 100 knots before the aircraft lifts off and climbs away. The A26 will reach more than 200 knots, or nearly 250 miles per hour, en route to the fire, where the bird dog will be waiting to direct us where to drop the load to prevent the fire spreading. The movie *Always* used the A26, but there the similarity to reality ended. The amphibious CL215s have a different role. They scoop up water by skimming over a lake in a continuous circuit, then dropping it over the fire for a cooling and holding action.

Lookout towers normally report fires. Spotters, usually university students on summer contract, man towers on hilltops overlooking areas where there is choice timber. The passage of a summer thunderstorm after a dry period will generally generate new fires. The person in the lookout tower takes a compass bearing on the first smoke, then radios the Forestry Dispatch. The dispatcher calls the base manager at the airport, who rings a loud bell to alert the crew. In about two minutes the old A26s start up, belching smoke and oil, then settle down to a powerful roar. The 'loaderman' puts the hose in the belly of our tanker as we taxi into the loading pit in succession. Each 26 takes on an 8,000-pound load of slurry, usually a phosphate-based chemical mixed with water as a transporting system. This is a long-term retardant that prevents the advance of the fire and remains in place, eventually adding to renewed growth as a fertilizer. When the slurry is loaded, it has to be dumped soon or it will gel and plug up the doors and delivery plumbing. We might not have a fire for a week, in which case we would have to dump the chemical on the ramp. It is too expensive to waste.

The crews will have been loafing about the crew room for hours, or maybe days, waiting for the bell to ring. We play chess, read, cook meals, swap lies, practice our golf swing, or play with our hobbies, to wait the dispatch bell. When it rings, everything is dropped and everyone runs to his aircraft.

Every morning we check our fuel, oil, radios, preset every switch and instrument. Even our seat harness is placed in the position so that it can be buckled quickly. Maps are folded, chocks removed and aircraft are ready to go. No one is allowed in our aircraft when we are on stand-by in case a switch or lever is moved. We don't have time to do a thorough cockpit check when the bell has rung. Fires spread very quickly on a hot day with low humidity and a fresh wind. Time is very important, and one tanker load across the head of a wildfire early will do more good than ten loads later in the day when the fire is running out of control. Worse still, it might rage towards an oil well, sawmill, a town, settlement or reserve.

Eight different groups are in operation in Alberta each year. In addition, numerous helicopters can be contracted in the case of a high fire risk period. The crews moved by helicopters have a variety of duties. They can carry 'rap' crews that rappel down ropes alongside the fire and set up bases and pumps. The choppers can also take the fire boss around to view the fire from the air so he can judge best how and where to position his resources and crews. To squelch smaller outbreaks 'spotting' in the wind from the main fire the helicopters, can also deploy a canvas bucket to pick up water from a nearby lake or stream. The 'skimmers' make water drops and keep it safe for the ground crews to work close to the flames. There are crews with hoses and shovels, and the tankers that drop chemical retardant ahead of the fire. All are important, of course, and all respect the other's contribution. As in a military action, however, rivalry exists. Everyone, to some degree, feels his talent is not fully utilized or appreciated. At least it gives us all something to whine about.

On my anniversary we are based at Edson on the Yellowhead Highway, a short drive through the beautiful foothills to Jasper in the Rocky Mountains. Our group, called Group 6, has four A26s and one bird dog. Bird dog pilot and group leader is Chris Chandler, a family man from Armstrong, B.C., who has been fighting forest fires for ten years. He is very efficient and clears the path for our tankers to make our drops. He flies the dummy run through the smoke to ensure there are no high trees or obstructions that the tankers might hit as they lumber down to less than 100 feet above the fire at minimum safe speed to drop their loads. The bird dog pilot is accompanied by a senior forestry official who decides where the load should go, how it will be dropped -- one door at a time, two doors in tandem, or as a salvo all in one spot. It depends on the characteristics of the fire and of the terrain.

Our Bird Dog Officer on July 22nd is Paul Rizolli from Lac la Biche, a 31-year veteran of the fire fighting business. As I write, Chris is chipping golf balls and Paul is resting under a tree - but both very close to their aircraft as the bird dog always goes to the fire first when the bell rings. They have the greatest responsibility for the success of our bombing missions.

Highly experienced pilots man our tankers. With 11 years in fire fighting and 28 in aviation, Pete Mitchell has flown jets and four engine tankers. He is at this moment preparing computer programmes for a golf club of which he is a co-owner. He lives in Calgary as a rich bachelor. Pete flies No. 36. Mike Hogan, from Tofino, B.C., flies No. 32. Mike came to Canada from Ireland as a boy and brought all the charm, wit and passion of a true Irishman. He learned to fly years ago in Calgary and enjoys the A26. In the winter, he flies tourists and fishermen in floatplanes on the west coast.

Butch Foster spent much of his flying career in the RCAF and has an impressive background on all the fighter jets currently in use by our military. He builds replica scale sized aircraft, which he flies for a hobby. His winter job is flying instructor with Mount Royal College in Calgary. Butch flies No. 12 and is positioned in the pit, meaning he will be the first loaded and gone when the bell goes. Butch has been doing this every summer since he left the RCAF thirteen years ago. The airport base manager is D'Arcy MacDonald, who started in Fire Control as a mixer at age 15, some 17 years ago. Our dispatcher, Larry Warren, has been in the Forestry Department for 18 years and is a real gentleman to work with, a logical practical leader.

Apart from the loneliness of living away from home, and having to cook our own grub or eat in restaurants, this is a great job. The pay is good; the camaraderie is great even after spending 93 days together from about breakfast until late evening every day, and never more than a few hundred feet apart. Even in flight, we all work together. It's a fraternity best understood by military men or people who have worked together in isolation.

I truly hate to leave A26 No. 31, known officially as C-GHCC. The venerable bomber was designed in 1942 and built in 1944. When this season ends, it will be taken over by a younger man, possibly our engineer, Tony Blake, an experienced pilot. Tony has been doing a top-notch job maintaining our aircraft in the field, doing ferry flights with other pilots in the dual controlled aircraft and deserves a seat next year. Tony's assistant, and the only other man on Group 6, is Hai Tran, an immigrant from Viet Nam. He has spent his entire working life in Canada with Airspray. He can beat me regularly at chess

A veteran of the war in Korea, No. 31 fought the fire wars in western Canada.

Airspray has a total of 40 pilots and 20 engineers spread around Alberta and the Yukon and the same people come back year after year. I'll miss that. I have watched sallow, beardless lads grow into mature family men and older men like Hy Baker, from Cardston, retire after many years of loyal service to the Alberta government. Hy Baker started his career putting out fires on foot, and on horseback with the equipment brought into the mountains on mules. Recently he retired to become an entrepreneur. Although we miss him, the cycle moves on. So be it.

Before the anniversary day ended, the bell rang and we were dispatched westward across the Continental Divide, which in this case is the B.C.-Alberta provincial boundary. The group passed magnificent Mount Robson towering above us to nearly 13,000 feet. Sightseeing over, we dropped on a fire in the Upper Fraser River Valley of British Columbia. This area is normally protected by the British Columbia Forest authority and served by Conair Aviation, but all their aircraft were busy elsewhere.

Some modern rangers and environmentalists now think that we tamper with nature too much. They feel that many forest fires should be left to burn and eventually be snuffed out by rain, that the forest renews itself naturally by fire. The large fire in Yellowstone National Park a few years ago, which was considered a national disaster at the time, is now considered an asset as the forest and wildlife have reproduced as never before.

And this is what Harry and I are leaving this year. Harry's group was at Fort McMurray and finished this week. Our group finishes after Labour Day. We will miss our summer family. Each province has a similar organization to protect their forests and like Alberta most of their crews are veterans.

The years have passed when I had a great compulsion to teach others flying. My sons are more current in modern aircraft than I. Technology and regulations change constantly. Students I have taught to fly have gone far ahead in equipment and salary, and many have retired already. The time has come to index the pictures in my albums (164 albums at last count) add up my log book totals (17 logbooks in all), but many, many pages are pictures of aircraft crews and anecdotes which made this book possible.

Dead reckoning, radio ranges, needle ball and airspeed are history. It's time to leave my basic flying lore and skills to younger, fresher pilots more highly skilled in computer science. To quote my late colleague Neil Armstrong: FLY SAFELY FRIENDS.

Memories of flying - Happy memories!